Augustin Privat-Deschanel, J. D. Everett

Elementary Treatise on natural Philosophy

Part I

Augustin Privat-Deschanel, J. D. Everett

Elementary Treatise on natural Philosophy
Part I

ISBN/EAN: 9783337076313

Printed in Europe, USA, Canada, Australia, Japan

Cover: Foto ©ninafisch / pixelio.de

More available books at **www.hansebooks.com**

ELEMENTARY TREATISE

ON

NATURAL PHILOSOPHY.

BY

A. PRIVAT DESCHANEL,

FORMERLY PROFESSOR OF PHYSICS IN THE LYCÉE LOUIS-LE-GRAND,
INSPECTOR OF THE ACADEMY OF PARIS.

TRANSLATED AND EDITED, WITH EXTENSIVE ADDITIONS,

By J. D. EVERETT, M.A., D.C.L., F.R.S.E.,

PROFESSOR OF NATURAL PHILOSOPHY IN THE QUEEN'S COLLEGE, BELFAST.

IN FOUR PARTS.

Part I.

MECHANICS, HYDROSTATICS, AND PNEUMATICS.

ILLUSTRATED BY

181 ENGRAVINGS ON WOOD, AND ONE COLOURED PLATE.

LONDON:

BLACKIE & SON, PATERNOSTER BUILDINGS, E.C.
GLASGOW AND EDINBURGH.

1872.

DESCHANEL'S NATURAL PHILOSOPHY.

By Professor J. D. EVERETT, D.C.L.

AUTHOR'S PREFACE.

THE importance of the study of Physics is now generally acknowledged. Besides the interest of curiosity which attaches to the observation of nature, the experimental method furnishes one of the most salutary exercises for the mind—constituting in this respect a fitting supplement to the study of the mathematical sciences. The method of deduction employed in these latter, while eminently adapted to form the habit of strict reasoning, scarcely affords any exercise for the critical faculty which plays so important a part in the physical sciences. In Physics we are called upon, not to deduce rigorous consequences from an absolute principle, but to ascend from the particular consequences which alone are known to the general principle from which they flow. In this operation there is no absolutely certain method of procedure, and even relative certainty can only be attained by a discussion which calls into profitable exercise all the faculties of the mind.

Be this as it may, physical science has now taken an important place in education, and plays a prominent part in the examinations for the different university degrees. The present treatise is intended for the assistance of young men preparing for these degrees; but I trust that it may also be read with profit by those persons who, merely for purposes of self-instruction, wish to acquire accurate knowledge of natural phenomena. Having for nearly twenty years been charged with the duty of teaching from the chair of Physics in one of the lyceums of Paris, I have been under the necessity of making continual efforts to overcome the inherent difficulties of this branch of study. I have endeavoured to turn to account the experience thus acquired in the preparation of this volume, and I shall be happy if I can thus contribute to advance the taste for a science which is at once useful and interesting.

For the convenience of candidates for the Bachelor's degree, I have appended to this treatise a number of problems, most of which have been taken from the examinations of the Faculty of Sciences of Paris or of the departments. With the same view I have made it my object to omit from the work none of the formulæ which are usually required for the solution of such questions. Beyond this point I have made very limited use of algebra. Though calculation is a precious and often indispensable auxiliary of physical science, the extent to which it can be advantageously employed varies greatly according to circumstances. There are in fact some phenomena which cannot be really understood without having recourse to measurement; but in a multitude of cases the explanation of phenomena can be rendered evident without resorting to numerical expression. In such cases calculation is of secondary importance, and may be said to be merely practical.

The physical sciences have of late years received very extensive developments. Facts have been multiplied indefinitely, and even theories have undergone great modifications. Hence arises considerable difficulty in selecting the most essential points and those which best represent the present state of science. I have done my best to cope with this difficulty, and I trust that the reader who attentively peruses my work, will be able to form a pretty accurate idea of the present position of physical science. I shall be happy in a second edition to avail myself of any observations which may be communicated to me on this or any other point.

TRANSLATOR'S PREFACE.

THE "TRAITÉ ÉLÉMENTAIRE DE PHYSIQUE" of Professor Deschanel, though only published in 1868, has already obtained a high reputation in France, and has been adopted by the Minister of Instruction as the text-book for Government Schools.

I did not consent to undertake the labour of translating and editing it till a careful examination had convinced me that it was better adapted to the requirements of my own class of Experimental Physics than any other work with which I was acquainted; and in executing the translation I have steadily kept this use in view, believing that I was thus adopting the surest means of meeting the wants of teachers generally.

The treatise of Professor Deschanel is remarkable for the vigour of its style, which specially commends it as a book for private reading. But its leading excellence, as compared with the best works at present in use, is the thoroughly rational character of the information which it presents. There is great danger in the present day lest science-teaching should degenerate into the accumulation of disconnected facts and unexplained formulæ, which burden the memory without cultivating the understanding. Professor Deschanel has been eminently successful in exhibiting facts in their mutual connection; and his applications of algebra are always judicious.

The peculiarly vigorous and idiomatic style of the original would be altogether unpresentable in English; and I have not hesitated in numerous instances to sacrifice exactness of translation to effective rendering, my object being to make the book as useful as possible to English readers. For the same reason I have not scrupled to suppress or modify any statement, whether historical or philosophical, which I deemed erroneous or defective. In some instances I have endeavoured to simplify the reasonings by which propositions are established or formulæ deduced.

As regards weights and measures, rough statements of quantity have generally been expressed in British units; but in many cases the numerical values given in the original, and belonging to the metrical system, have been retained, with or without their English equivalents; as it is desirable that all students of science should familiarize themselves with a system of weights and measures which affords peculiar facilities for scientific calculation, and is extensively employed by scientific men of all countries. For convenience of reference, a complete table of metrical and British equivalents has been annexed.

The additions, which have been very extensive, relate either to subjects generally considered essential in this country to a treatise on Natural Philosophy, or to topics which have in recent years occupied an important place in physical discussions, though as yet but little known to the general public.

The sections distinguished by a letter appended to a number are all new; as also are all foot-notes, except those which are signed with the Author's initial "D."

In many instances the new matter is so interwoven with the old that it could not conveniently be indicated; and I have aimed at giving unity to the book rather than at preserving careful distinctions of authorship.

Comparison with the original will however be easy, as the numbering of the original sections has been almost invariably followed.

The chief additions in Part I. (Chap. i.–xviii.) have been under the heads of Dynamics, Capillarity, and the Barometer. The chapter on Hydrometers has also been recast.

ADVERTISEMENT TO REPRINT OF PART I.

The first impression of Part I. having been exhausted, opportunity has been taken, in the present reprint, to extend the Table of Contents, and to make a few unimportant corrections and additions in the body of the work.

CONTENTS—PART I.

—

FRENCH AND ENGLISH MEASURES.

A DECIMETRE DIVIDED INTO CENTIMETRES AND MILLIMETRES.

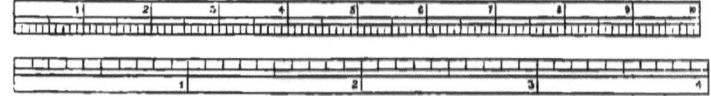

INCHES AND TENTHS.

TABLE FOR THE CONVERSION OF FRENCH INTO ENGLISH MEASURES.

Measures of Length.

1 Millimetre	=	·03937079 inch, or about $\frac{1}{25}$ inch.
1 Centimetre	=	·3937079 inch.
1 Decimetre	=	3·937079 inches.
1 Metre	=	39·37079 inches, or 3·2809 feet nearly.
1 Kilometre	=	39370·79 inches, or 1093·6 yards nearly.

Measures of Area.

1 sq. millimetre	=	·00155006 sq. inch.
1 sq. centimetre	=	·155006 sq. inch.
1 sq. decimetre	=	15·5006 sq. inches.
1 sq. metre	=	1550·06 sq. inches, or 10·7643 sq. feet.

Measures of Volume.

1 cubic centimetre	=	·0610271 cubic inch.
1 cubic decimetre	=	61·0271 cubic inches.
1 cubic metre	=	61027·1 cubic inches, or 35·3166 cubic feet.

The Litre (used for liquids) is the same as the cubic decimetre, and is equal to 1·76172 imperial pint, or ·220215 gallon.

Measures of Weight (or Mass).

1 milligramme	=	·015432349 grain.
1 centigramme	=	·15432349 grain.
1 decigramme	=	1·5432349 grain.
1 gramme	=	15·432349 grains.
1 kilogramme	=	15432·349 grains, or 2·20462125 lbs. avoir.

Measures involving reference to two units.

1 gramme per sq. centimetre	=	2·048098 lbs per sq. foot.
1 kilogramme per sq. metre	=	·2048098 " "
1 kilogramme per sq. millimetre	=	204809·8 " "
1 kilogrammetre	=	7·23314 foot-pounds.

1 force de cheval = 75 kilogrammetres per second, or 542½ foot-pounds per second nearly, whereas 1 horse-power (English) = 550 foot-pounds per second.

TABLE FOR THE CONVERSION OF ENGLISH INTO FRENCH MEASURES.

Measures of Length.

1 inch = 25·39954 millimetres.
1 foot = ·30479449 metre.
1 yard = ·91438347 metre.
1 mile = 1·60932 kilometre.

Measures of Area.

1 sq. inch = 645·137 sq. millimetres.
1 sq. foot = ·0928997 sq. metre.
1 sq. yard = ·8360973 sq. metre.
1 sq. mile = 2·589895 sq. kilometres.

Solid Measures.

1 cubic inch = 16386·6 cubic millimetres.
1 cubic foot = ·0283153 cubic metre.
1 cubic yard = ·7645131 cubic metre.

Measures of Capacity.

1 pint = ·5676275 litre.
1 gallon = 4·54102 litres.
1 bushel = 36·32816 litres.

Measures of Weight.

1 grain = ·064799 gramme.
1 oz. avoir. = 28·3496 grammes.
1 lb. avoir. = ·453593 kilogramme.
1 ton = 1·01605 tonne = 1016·05 kilos.

Measures involving reference to two units.

1 lb. per sq. foot = 4·88261 kilos. per sq. metre.
1 lb. per sq. inch = ·0703095 kilos. per sq. centimetre.
1 foot-pound = ·138253 kilogrammetre.

c

ELEMENTARY TREATISE

ON

NATURAL PHILOSOPHY.

CHAPTER I.

PRELIMINARY NOTIONS.

1. **Origin of Natural Philosophy.**—The object of Natural Philosophy or Physics (φύσις, nature) is the study of the material world, including the phenomena which it presents to us, the laws which govern them, and the applications which can be made of them to our various wants.

In its widest sense, the study of physics must be traced back to the origin of the human race; for ever since man came into being, he must necessarily have been struck by the spectacle of the heavens and the continually changing aspect of terrestrial phenomena. But isolated and vague observations, and the barren admiration of phenomena which provoke attention or excite curiosity, do not constitute science; this can only exist where there is a mass of accurate knowledge in which the facts are related to each other and studied in connection with the causes which produce them. This process of co-ordination is only possible after a considerable collection of facts has been accumulated; but it then becomes inevitable, from the very constitution of the human mind. Thus, in examining the history of the nations among whom we place the cradle of our civilization, we find constant efforts of philosophers to explain the mechanism of the external world,—to bring all the facts which nature presents to us under one theory—one system. The Greek philosophers, especially, who appear to have borrowed the greater part of their physical knowledge from the Egyptian priests, have left us different systems, by the aid of which they profess to explain all natural phenomena. Thus Thales, the most celebrated of the seven wise men of Greece

1

(640 B.C.), made water a universal principle, which nourishes at once the sun, the earth, and the planets. Plato (398 B.C.) assumed two distinct principles, matter and form, which by their combination give birth to five elements—earth, water, fire, air, and ether. According to Anaximander, there is but one principle, the infinite, which gives birth to all bodies. According to Anaxagoras, air is the sovereign of nature. We need not stay to examine the exact meaning of these propositions, which, taken in their literal sense, appear at the present day sufficiently unintelligible. While acknowledging that these illustrious philosophers knew and taught some important facts of general physics, we are bound to remark that in the elaboration of their systems experiment played no part; that observation itself only held a secondary place; and that their theories were veritable *à priori* conceptions, to which facts had to be accommodated. Hence there is nothing in their works approaching the experimental method which serves as the foundation of modern physics. Some faint foreshadowings of this method may be traced in the works of Aristotle (383 B.C.), who was a disciple of Plato, but far superior to his master in scientific genius, besides being an eminent naturalist, and author of a history of animals, which alone would constitute an imperishable monument to his memory. Thus, to investigate the weight of air, he had recourse to a direct experiment, which consisted in weighing a skin empty and inflated. Finding no difference in the weights, he concluded wrongly that air is destitute of weight, and was thus led in his attempts at the explanation of certain phenomena to the famous principle that nature abhors a vacuum, which was universally admitted down to the time of Galileo.

It was especially in the hands of Archimedes (287 B.C.), and the philosophers of the school of Alexandria, who may be regarded as his successors, that the method of scientific observation took a distinct form, and led to important results. Every one has heard of the admirable discoveries of Archimedes respecting the theory of the lever, the determination of centres of gravity, and the measurement of specific gravities by means of the principle which bears his name; discoveries founded upon experiments which were doubtless not very accurate, but were regarded by him as necessary in order to furnish a solid basis for his investigations. After him Hipparchus (140 B.C.), by means of persevering observations, methodically directed, changed the face of astronomy, and arrived at brilliant discoveries, among which the most notable was that of the precession of the equinoxes. At a later period, Ctesibius, Hero, Posidonius, &c., following in the traces of their illustrious predecessors, advanced the boundaries of the exact knowledge already acquired, and originated several inventions displaying more or less ingenuity. To the first of these philosophers the invention of pumps appears to be due, and the fountain of Hero still finds a place in all collections of physical apparatus.

Among philosophers belonging more or less directly to the school of Alexandria, who have enriched science by important discoveries, we will only mention Plutarch, who is said to have discovered the refraction of light in its passage from air into water; and Ptolemy, the author of various works on optics and celestial physics, which constitute a better title to glory than the astronomical system which bears his name, a system which only served to retard the progress of science.

We will carry this historical review no further, but will content ourselves with remarking that, starting from the seventh century, the period of the conquest of Alexandria by the Arabs, and the burning of its celebrated library, until the time of Galileo (1564), science may be said to have been stationary. Still, some discoveries of importance belong to this period; for example, that of the mariner's compass, which was known from the thirteenth century. Shortly before Galileo, the thermometer, the microscope, and telescope were invented; but it is unquestionably to this distinguished philosopher that we owe the true scientific method—the method of experiment. His treatises upon falling bodies, the pendulum, &c., furnish admirable examples of the manner in which the physical investigator should interrogate Nature by the aid of experiment. It was by the introduction of this method that physical science became finally disentangled from the prejudices and à priori assumptions which had hitherto impeded its progress.

At the present day, after numberless discoveries which have introduced most material changes in our social condition, physical science has attained a very high degree of perfection. It is to the experimental method that we owe this result, and it is by remaining true to this method that we must hope to achieve fresh progress.

2. The Experimental Method.—The experimental method can easily be described in general terms: it consists in observing facts instead of trying to divine them; in carefully examining what really happens, and not in reasoning as to what ought to happen.

It is therefore entirely independent of metaphysics, which has always proved a false ally; in fact, as long as the dominion of metaphysics lasted, science continued to run in the old ruts, which it did not leave till, thanks to the teaching of Bacon and Galileo, the conviction became established that there is no way of arriving at physical truths but by the help of observation and experiment.

The experimental method is usually called by logicians the method of *observation* and *induction*. From the observation of particular facts it ascends to the general law which embraces them; being very different in this respect from the method of *deduction* employed in mathematics, in which we always descend from a certain and absolute principle to the different consequences which flow from it. Let us enter into some details upon this point.

3. **Phenomena—Physical Law.**—A phenomenon is any change that takes place in the condition of a body; the fall of a stone, the flowing of water, the melting of lead, the combustion of wood, for example, are phenomena. When we study the characteristics which belong to phenomena of the same class, we soon perceive that the various circumstances of their production have a mutual dependence, so that if one of them varies, the others undergo a corresponding variation. The expression of this connection constitutes a physical law.

Sometimes the law appears of itself and without difficulty, by means of observation alone. Such, for example, is the following: *All bodies left to themselves fall to the surface of the earth.* But more frequently the law is disguised by disturbing causes, whose influence should, as far as possible, be eliminated. This elimination is the object of experiment. Experiment differs from observation in this respect,—that the phenomenon is produced under conditions previously determined and regulated by the experimenter. If we wish to know, for example, what are the velocities which gravity produces in different bodies falling freely, we must not let them fall in air, because this fluid retards their movement, and that in unequal degrees for different bodies; we must operate *in vacuo*, and thus we arrive at the law, which observation alone could never have discovered, that *gravity produces the same velocity in all bodies.* It will be readily understood then that the art of experimenting, that is, of regulating the special conditions under which phenomena shall take place, and of measuring their constituent elements, is absolutely necessary to the physical investigator; and that a genius for physical science mainly consists in the possession of this aptitude in a more or less eminent degree.

We may remark that when the general law of a class of phenomena is known, the expression of this law is often called the physical cause of the particular phenomena which it includes. A phenomenon is said to be explained or accounted for, or traced to its cause, when we show that it is contained in the enunciation of a known law.

Thus, when we have once laid down the principle that the volumes of gases under different pressures vary inversely as these pressures, we are in a position to explain a crowd of facts depending on the action of a gas whose volume and pressure vary simultaneously.

When the law of observed phenomena admits of numerical statement, calculation becomes a valuable instrument for making known all its consequences, and the experimental verification of these consequences constitutes a confirmation of the physical law itself. In this way mathematical methods become powerful auxiliaries to physical science.

4. Physical Theory.—The enunciation of any one physical law, and the rational development of its consequences, constitute a partial physical theory. The assemblage of all the laws which belong to one class of phenomena, forms a more general physical theory; but it will be readily understood that these different laws may be merely corollaries of a single law.

The discovery of this single law, when it exists, marks a decided step in the progress of physical science. Thus Newton traced to the single law of gravitation all the movements of our planetary system, as well as those of bodies which fall to the surface of the earth.

In like manner the different partial theories of optics are rigorous consequences of the properties attributed to a fluid called *ether*, with which we suppose space to be filled, and whose vibrations serve for the propagation of light and heat.

This work of synthesis, however, has as yet made little progress, though these last few years have been marked by very successful efforts in this direction; but it should be considered as the true object of physical science in general, and the highest generalization will have been attained when it has been demonstrated that all the physical agents which have hitherto been regarded as distinct, are merely transformations of one and the same primordial agent.

5. Divisions of Physical Science.—Physical or natural science in general comprises the aggregate of all the phenomena of the external world; but the accumulation of discoveries in different parts of this mighty whole has necessitated the division of it into several branches, which at present constitute distinct sciences.

Natural History comprises all those facts which have reference to the different beings, organic or inorganic, which are found upon the surface of the globe; it is further subdivided into several parts. *Zoology* is occupied with the organization and habits of animals, with their regular classification, and with all the phenomena connected with their development and reproduction. *Botany* treats of the same questions with respect to vegetables. *Mineralogy* has for its object the description and methodical classification of the different inorganic bodies (minerals) which nature presents to us; the knowledge of the peculiar characteristics which serve to distinguish them from one another; and the enumeration of their principal properties as well as of the various applications that can be made of them.

Geology is the history of the earth; it recounts the different revolutions which have modified the surface of the globe and finally brought about its present configuration, the arrangement and nature of the rocks that enter into its composition, and the description of those ancient animals and vegetables whose fossil remains are still in existence, belonging in many cases to types which have since become extinct. It is the basis of the art of the mining engineer, and enables him to follow a regular method in searching for the various metals or combustible substances which are hid in the depths of the earth, and which we employ to satisfy our various requirements.

Astronomy is occupied with the laws of the movements of the heavenly bodies; thanks to the perfection to which our measuring instruments have been brought, to the progress of mathematical science, and to the discovery of the universal law of gravitation, astronomy has arrived at such a degree of perfection that it may be classed among the exact sciences.

Besides natural history and astronomy, there is room further for distinguishing physics from chemistry. This latter science, in fact, has for its object the study of phenomena in which the essential character of materials seems to be changed; phenomena in which matter seems to be destroyed, or at least metamorphosed. If we take a piece of sulphur and heat it, it will melt; if we rub it with a piece of wool, it will acquire the power of attracting light bodies, and will present the peculiar and curious properties which are characteristic of electrical excitation; but the sulphur will not have lost its proper nature, and when the different influences to which it has been submitted cease to act, it will resume all its original characteristics. The sulphur under these circumstances has displayed *physical* phenomena. If, on the other hand, we place this same body in a fire, we shall see it burn with a blue flame; at the end of some time it will have entirely disappeared, or at least will have been transformed into a gaseous substance which is dissipated with the other products of combustion. In this case the sulphur has ceased to exist as sulphur; a *chemical* phenomenon has taken place.

In a more restricted sense, then, physics or natural philosophy is understood as embracing the study of all the phenomena of the material world except those which consist in the action of vital forces or of chemical affinities. It is in this restricted sense that physics forms the subject of the present treatise. We may remark, however, that the two kinds of phenomena are often produced by the same causes, and that each is frequently the necessary consequence of the other. Thus in heating a body we render it better adapted to undergo chemical transformations; and, on the other hand, such transformations often produce a great quantity of heat. Physics and chemistry, though pursuing different ends, should yet afford each other mutual assistance. For example, our ideas of electricity would be very imperfect without a knowledge of the curious and often useful chemical phenomena which it is capable of producing.

A similar remark may also be made with regard to all the other branches of natural science. How, for example, can we separate mineralogy and chemistry, when it so often happens that the only means of recognizing a mineral is by making a chemical analysis of it? and when, on the other hand, a complete description of the substances which the chemist produces in his laboratory, must necessarily include an account of their external characteristics, such as their crystalline form, which specially belongs to the province of mineralogy?

To take another instance. Can we draw a sharp line of demarcation between zoology and botany on the one side, and physical science and chemistry on the other? Does not the tissue of organic beings undergo various chemical reactions which are a necessary accompaniment of vital phenomena? Do not physical agents in their turn produce phenomena of such a nature as completely to embarrass the physiologist and the physician, unless they are armed with a knowledge of the laws which regulate the action of these agents upon inorganic bodies?

Finally, though astronomy may seem to form a totally distinct science, consisting of the geometry and mechanics of the movements of the heavenly bodies, must it not avail itself of all the resources of physical science if it would arrive at any rational conjectures respecting their constitution?   We may say, then, that all the parts of natural science are interwoven together; they form one connected whole, and the division into distinct sciences has simply arisen from the vastness of the subject, which renders it impossible for any one mind adequately to follow the development of its various branches.

CHAPTER II.

6. Principle of Inertia.—The fundamental principle of physics is the inertia of matter. Inertia does not consist in the inactivity of material particles, nor in the impossibility of changes being produced in their states of rest or motion by their mutual action; for a glance at nature is sufficient to show that repose nowhere exists, and that motion changes in an endless variety of ways. The principle of inertia is an abstract principle which must be considered as applicable to a single isolated particle. It may be enounced in the following terms:—

An isolated material point cannot change its state, whether of rest or motion. That is to say, if it be at rest it will remain at rest; if it be in motion it will continue to move in the same direction and with the same velocity.

If, then, we see a material point which was at rest begin to move, or if we observe any change in the motion of a point, we say that it has been acted on by a *force*.

Without entering upon the very obscure subject of the intimate nature of forces—without seeking to know whether they form an essential part of bodies or have a separate existence, but only regarding them in the effects which they produce, we may define them in the following manner:—

A force is any cause which tends to urge a material point in a definite direction with a definite velocity.[1]

7. Manifestations of Inertia.—The principle of inertia, as above enounced, does not admit of direct experimental verification; for we cannot observe a material point, which is a mere abstraction;

---

[1] The words *with a definite velocity* only imperfectly express the idea intended to be conveyed. The correct phrase would be *with a definite acceleration.* See Chap. v.

still less an isolated material point. The principle of inertia is one of those ultimate and abstract principles which presented themselves to the minds of the founders of the science of mechanics—of Newton especially—as the key and reason of the manifold and complex characters of external phenomena. But if it is impossible to verify the principle of inertia directly, it is easy to show its influence in external phenomena, this influence reducing itself evidently to the tendency of bodies to continue in their state of rest or motion.

The tendency to continue in a state of rest is manifest to the most superficial observation. The tendency to continue in a state of uniform motion can be clearly understood from an attentive study of facts. If, for example, we make a pendulum oscillate, the amplitude of the oscillations decreases more and more; and ends, after a longer or shorter time, by becoming nothing. This is because the pendulum experiences resistance from the air, due to the successive displacement of the particles of this fluid; and because the axis of suspension rubs on its supports. These two circumstances combine to produce a diminution in the velocity of the apparatus until it is completely annihilated. If the friction at the point of suspension is diminished by suitable means, and the apparatus is made to oscillate *in vacuo*, the duration of the motion will be immensely increased.

Analogy evidently indicates that if it were possible to suppress entirely these two causes of the destruction of the pendulum's velocity, its motion would continue for an indefinite time unchanged.

This tendency to continue in motion is the cause of the effects which are produced when a carriage or railway train is suddenly stopped. The passengers are thrown in the direction of the motion, in virtue of the velocity which they possessed at the moment when the stoppage occurred. If it were possible to find a brake sufficiently powerful to stop a train suddenly at full speed, the effects of such a stoppage would be identical with those which would result from collision with another train of the same weight coming in a contrary direction with equal velocity.

Inertia is also the cause of the severe falls which are often received in alighting incautiously from a carriage in motion; all the particles of the body have, in fact, a forward motion, and the feet alone being reduced to rest, the upper portion of the body continues to move, and is thus thrown forward.

When we fix the head of a hammer on the handle by striking the end of the handle on the ground, we utilize the inertia of matter.

In fact, at the moment of the shock, and of the stoppage which results, the head continues to move, and ends after some blows by becoming firmly fixed.

8. Mechanics.—All physical phenomena fundamentally consist in motions; but these motions are in many cases too minute to admit of direct observation, and are only inferred from their effects. Thus when a solid body is heated and melted, it is certain that the liquid state results from a particular displacement of the molecules, and perhaps also from a change of their form—that is to say, from circumstances which are reducible to motions; but the liquid body thus formed has acquired peculiar properties, which form a subject of study in themselves apart from the motions to which they are due.

When motions are considered in themselves, according to their geometrical relations, and in connection with the forces which produce them, they form the subject of the science of mechanics, which must be regarded as an indispensable introduction to physics. We shall give in this chapter enunciations and illustrations of some fundamental propositions, referring the reader to special treatises on this subject for fuller information.

9. Elements required to specify a Force.—The material point submitted to the action of force is called the *point of application of the force*. It tends, in virtue of this action, to move in a certain direction, which is called the *direction of the force*, and which can be represented geometrically by a straight line drawn from the material point. It is obvious also that the force must act with some definite intensity, which is different in different cases. This intensity may manifest itself, for example, by a greater or less velocity of the point, a greater velocity corresponding to a greater force.

When two forces separately applied to the same point at rest give it the same motion, they may be called equal. The union of a number of equal forces gives a force which is a corresponding multiple of one of them, and thus the intensities of forces can be numerically compared. Forces then can be represented either by numbers or by lines; in the latter case a certain length (as an inch) being taken to represent a certain force

Fig. 1.

(as the weight of a pound). It is usual to indicate the direction of a force by a line AF with which the direction of the force coincides, and to lay off on this a length AB representing (on the scale chosen) the intensity of the force.

For accuracy, it is to be observed that the pound, ounce, and other units of weight are essentially units of mass, not of force. In order to render them available as accurate units of force, the *locality* must be specified, inasmuch as the force requisite to support a pound of matter is different in different localities, being for example greater at the poles of the earth than at the equator by about 1 part in 190.

10. Resultant.—When a material point or a system of points is urged by a certain number of forces, it will be readily understood that a single force of determinate magnitude, and applied at a suitable point, may be capable of producing the same effect as all the given forces acting together. This single force is called the *resultant* of the given forces, and they are called its *components*.

Thus, for example, a vessel descending a river, whether propelled by steam or wind, provided its motion be rectilinear, is really urged forward by a great number of forces applied at different points; but it is evident that a single force of proper magnitude and line of action would produce the same effect.

It is not every system of forces that has a resultant; but, in the case of those which have, it is very important to determine its magnitude and position, for the study of the body's motion will thus be evidently simplified. The following is an important case in which this determination is easily made.

11. Parallelogram of Forces.—*If a material point A is acted on by two forces represented in magnitude and direction by AB and AC, there is a resultant, which is exactly represented by the diagonal AD of the parallelogram of which AB and AC are sides.*

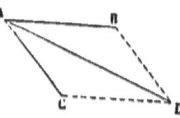

Fig. 2—Parallelogram of Forces.

This proposition can be verified experimentally by the aid of the following apparatus due to Gravesande. ABDC (Fig. 3) is a parallelogram jointed at its four corners. To the points B and C cords are fixed, which, passing over the pulleys M and N. support at their extremities weights P and P', of 90 and 60 ounces respectively.

The lengths of the sides AB and AC are themselves proportional to the numbers 90 and 60. To the corner A is attached a weight P'' of 120 ounces. In these circumstances, the parallelogram will take a position of equilibrium, in which the cords attached to B and C will be found to form prolongations of the sides AB, AC,

and the diagonal AD will be vertical.   But the forces P and P′ have
a resultant acting vertically at A, since their resultant must be equal
and opposite to the weight P″ which balances them.   The diagonal

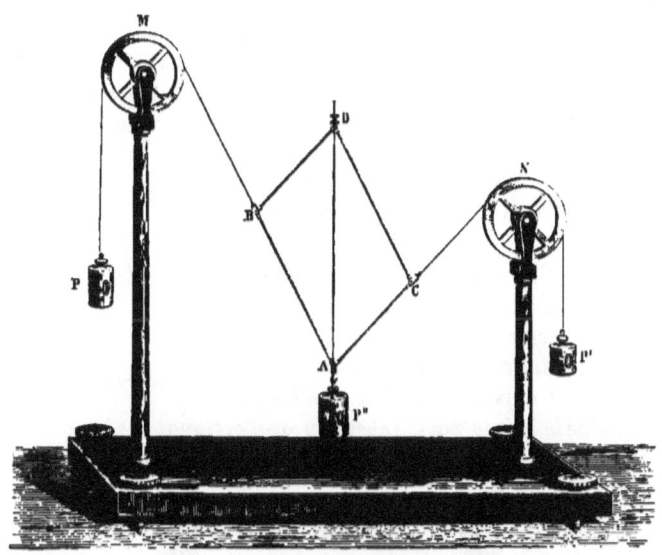

Fig. 3.—Gravesande's Apparatus.

AD therefore agrees with the resultant in direction; and if this
diagonal is measured, its length will be found to be 120 on the same
scale on which the lengths of AB and AC are 90 and 60.

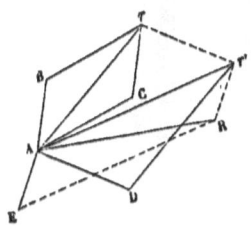

Fig. 4.—Composition of any Number
of Forces.

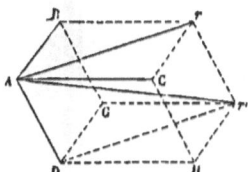

Fig. 5.—Parallelopiped of
Forces.

12. **Composition of Forces.**—Knowing how to find the resultant of
two forces, that is to say, to *compound* two forces, applied to the
same point, it is easy to compound any number.

Let there be, for example (Fig. 4), four forces applied to the material point A. We may compound first the force AB with AC, which gives the resultant Ar; this, compounded with AD, gives a second partial resultant Ar', which, compounded with the fourth force, gives the complete resultant AR.

In the particular case of three forces (Fig. 5), it is easily seen that the resultant Ar' is the same thing as the diagonal of the parallelopiped constructed on the lines AB, AC, AD which represent the three forces. In the figure, the parallelopiped has been completed to render this evident, but the construction amounts, as in the preceding case, to compounding AB with AC, and their resultant Ar with AD.

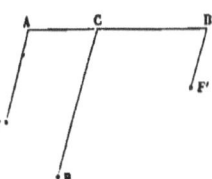

13. Composition of Parallel Forces.—*When two parallel forces* F *and* F' *are applied at the two extremities of a straight line, they have a resultant* R *equal to their sum, and acting at a point* C *which divides the straight line* AB *into parts inversely proportional to the forces.*

Fig. 6.—Parallel Forces.

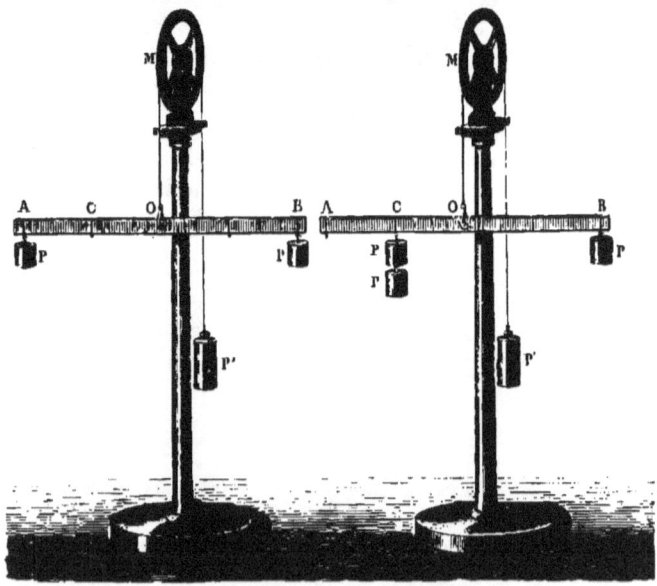

Fig. 7.—Composition of Parallel Forces.

If, for example, the two forces F and F' are equal, the point C will be at the middle of AB; if the force F is double of F', the segment CA will be equal to half of CB.

This proposition can be verified by the aid of the following apparatus called the *arithmetical lever* (Fig. 7).

The lever AB supports two equal weights P at its extremities; it is suspended at its middle by a cord which, passing over the pulley M, sustains a weight P'. It will be found that, when the weight P' has a certain value, the lever is in equilibrium; whence it follows that the two weights P and the weights of the different parts of the lever, which we may suppose distributed two and two at equal distances from the middle point, have a resultant acting at this point, equal and opposite to the force P'. It will also be found that P' is equal to the sum of the two weights P together with the weight of the lever.

In the case of the second figure, a single weight P is placed at one of the extremities B, whilst two equal weights are suspended at the middle point C of the second half of the lever. It will be found that there is still equilibrium, provided that the weight P' is the sum of the three weights P and the weight of the lever.

To interpret this result, we may remark that the lever being balanced directly by an equal portion of P', we may neglect its weight; there remain then only two forces, of which one, that on the left, is double of the other. Now the resultant evidently passes through the point of suspension O, which is exactly twice as far from B as from C.

14. When the parallel forces F and F' act in opposite directions, there is still a resultant parallel to the components, acting in the same direction as the greater, and equal to their difference F—F'. Moreover its point of application C is so placed that the distances CA and CB are inversely proportional to the forces, a result analogous to that which holds when the forces act in the same direction.

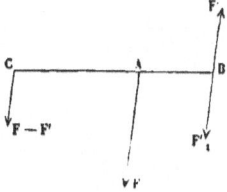

Fig. 8.—Parallel forces in Opposite Directions.

We see, as a consequence of this proposition, that if the two parallel and opposite forces differ little from one another, their resultant has a very small value, but its point of application is very remote. In the particular case in which the two forces are equal, the rule of composition is absolutely inapplicable. Such a system,

consisting of two equal and parallel forces acting in opposite directions, is called a *couple.* It cannot be equilibrated or replaced by
a single force, but obviously tends to produce a motion of rotation.
Now in nature we frequently see bodies possessing at once a motion
of translation and a motion of rotation. We may assume that the
translation has been produced by a force and the rotation by a
couple; this latter then presents itself as a sort of *natural element*
in mechanics. The idea of couples originated with the geometrician
Poinsot, and has greatly simplified many mechanical problems.

The perpendicular distance between the lines of action of the two
equal forces which constitute a couple is called the *arm* of the couple.
The product of one of the two equal forces by the arm is called the
moment of the couple, and is the measure of the power of the couple
to produce rotation. It is proved in treatises on mechanics that
two couples acting on a body and tending to turn it in opposite
directions will equilibrate each other if their moments are equal,
even though they be applied at different parts of the body. Two or
more couples acting on a body and tending to turn it in the same
direction, may be replaced by a single couple whose moment is the
sum of their moments; and any number of couples acting on a body
and tending to turn it in any directions whatever, are always, except
when they are in equilibrium, equivalent to
a single couple.

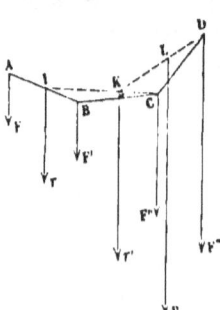

Fig. 9. — Composition of any
Number of Parallel Forces.

**15. Composition of any Number of Parallel
Forces.** To compound a given number of
parallel forces, F, F', F", F"', we may first
compound the first with the second, which
gives a partial resultant r; this compounded
with the third force F", gives a second resultant r', which combined with F"', gives the
complete resultant R. It is clear that this
procedure is applicable to any number of
forces, and that the resultant is always equal
to the sum of all the forces. As to its point of
application, we may remark, that the point of application of the
first partial resultant is obtained by dividing AB into two parts
AI and BI inversely proportional to F and F'. In like manner
the point of application K of r' is obtained by dividing IC into parts
IK and KC, inversely proportional to r and F"; and lastly, the point
of application L of the complete resultant is obtained by performing

an analogous operation on the line KD. Now it will be remarked that this series of constructions is independent of the absolute direction of the forces, and only supposes that they are parallel; if then the forces were turned about their points of application in such a manner as still to remain parallel to one another, their resultant would still pass through the same point L. This point is on this account called the *centre of parallel forces*.

16. Resolution of Forces.—As any number of forces can be compounded into a single force, so a given force can be (decompounded or) resolved into two or any number of forces which would produce the same effect. Thus, for example, the force AD (Fig. 2) can be replaced by the two forces AB and AC, inasmuch as it is their resultant. It is obvious that, to resolve a force applied to a material point into two others having given directions, it is necessary to draw, through the extremity D of the line which represents the given force, lines parallel to the given directions. They determine by their intersections the points B and C, and consequently the magnitudes AB and AC of the two forces which can be substituted for the given force.

The composition of several forces into a single force constitutes an obvious simplification. As to resolution (or decomposition), its utility is not so obvious. It may appear at first sight to be a complication. Such however is not always the case, and in the course of this treatise we shall find it of continual use. It will be readily conceived, in fact, that when a force is applied to a point whose movements are constrained by guides or physical connections, the effect which the force can produce is not easily perceived. But it is sometimes possible, by resolution, to replace it by components, of which some are destroyed by the conditions of constraint, and the others can act in a manner directly appreciable.

We will confine ourselves to a single example of this, namely, the explanation of the opposite courses which can be taken by two sailing boats with the same wind.

Consider, for example, the case of the first figure. The wind, blowing in the direction Vm, tends to exert a force which can be resolved at the point m into two components, one of them tt' tangential to the sail, which has no effect, the other mn perpendicular to the sail. This latter tends to urge the boat obliquely towards the left of the figure. But as the boat can be moved much more easily in the direction of its length than in any other direction, we may resolve this latter force into two components: one perpendicular to the length, which has little effect; the other in the direction of the length, which produces the forward movement of the boat in the direction of the arrow placed below the figure.

Fig. 10.—Resolution of Pressure of Wind on Sails.

In the second figure it will be seen that the same methods of resolution lead to an opposite result, and that the direction of motion is opposite, though the direction of the wind is the same. In reality, the two motions are not directly opposite, because the action of the component perpendicular to the length of the boat cannot be altogether neglected, but produces what is called leeway.

17. Work done by a Force.—In the different operations to which forces are applied, such as the raising of burdens, the compressing, piercing, and pulverizing of solid bodies, it is clear that it is always necessary to overcome a certain resistance and produce a certain displacement. Hence we are led to a special mechanical element, involving the joint consideration of force and the displacement of its point of application. This element is called *work*.

The work done by a force is the product of the force into the displacement which it produces in its point of application.

In this definition the force is supposed to be constant, and the motion of the point of application is supposed to coincide with the direction of the force.

In stating the work done by a machine it is usual to take as unit of work the foot-pound; that is to say, the work done in raising a pound through a height of a foot.

The notion of work is itself independent of time; but it is evident that in practice it is advantageous for a machine to employ little time in producing a given amount of work.

The unit employed for expressing the rate of working of a machine is the *horse-power*, and denotes 33,000 foot-pounds of work done per minute or 550 foot-pounds per second. Thus a machine which can raise 12 tons through a height of 10 feet in 2 minutes is a machine of rather more than 4 horse-power; since it does $12 \times 2240 \times 10 = 268,800$ foot-pounds in 2 minutes, or 134,400 in 1 minute; and 134,400 is rather more than 4 times 33,000.

17A. The above definition of *work* is only applicable to the case in which the displacement of the point of application of the force is in a direction precisely coincident with the direction of the force. It is often necessary to consider cases where (owing perhaps to circumstances of constraint, or to the action of other forces besides the one considered, or to previous motion) the point of application of a force moves in a direction oblique to that of the force. In this case the force may be resolved into two components, of which one is perpendicular and the other either coincident with or directly opposite to the direction of displacement. The former of these components is to be neglected in estimating the work done by the force; and the product of the latter component by the displacement is the work done *by* or *against* the force according as the direction of this component *coincides with* or is *opposite to* that of the displacement.[1]

The necessity of having a name to denote the idea thus defined is obvious from the following proposition, which is called the *principle of work.*

Every machine may be regarded as an instrument for transmitting work; and if we neglect friction, we may assert that *the work thus transmitted is unaltered in amount.* If, for example, the machine is driven by forces applied at points A_1, A_2, &c., and if the machine overcomes resistances at points B_1, B_2, &c., then the whole work done by the forces at A_1, A_2, &c., estimated according to the fore-

---

[1] Or the work done by a force is the product of the force by the projection of the displacement of its point of application on the direction of the force, or is the continued product of force, displacement, and cosine of included angle. The three definitions are obviously equivalent. Work done *against* a force is to be regarded as *negative work done by* the force.

going definition, will be precisely equal to the whole work done against the resistances at B_1, B_2, similarly estimated.

The numerous vain schemes for producing perpetual motion are founded on ignorance of this law. They are attempts to make work increase in its transmission through a machine.

Practically, work is always diminished in its transmission through a machine, owing to friction. The work thus lost leaves an equivalent in the shape of heat (see Chap. xxxii.)

CHAPTER III.

18. Different States of Matter.—As the object of physics is the study of the general properties of bodies, it is necessary for us to form some idea of the constitution of the different kinds of matter. Matter presents itself in three different states: the solid, liquid, and gaseous. Solid bodies are characterized by a kind of invariability of form; that is to say, their form cannot be changed without an effort, more or less considerable. Hence a solid body forms a firmly connected whole, so that the movement of one of its parts produces motion in the rest.

Liquids, on the contrary, appear to be formed of particles which are independent of each other and can obey individually the action of the forces which urge them, being able to slide past each other with the greatest facility. From this property the name *fluids*, by which, in common with gases, they are often designated, is derived (*fluere*, to flow). This also is the reason why a liquid moulds itself to the form of the containing vessel. Liquidity, consisting essentially in the perfect mobility of the constituent parts of a body, may evidently be met with in different degrees of perfection. Thus sulphuric ether and alcohol are more perfectly liquid than water; water itself is more liquid than oil, and so on. *Viscosity* is a name used to denote the want of independence between the particles of a liquid, which establishes a kind of intermediate state between these bodies and solids. Thus we may say that there is an insensible passage from liquids more or less perfect to viscous liquids, from these to plastic substances such as putty or moist clay, and from these last to solid bodies.

Gaseous bodies, of which the atmosphere offers us an example, are formed, like liquids, of independent particles: but these particles appear to be in a continual state of repulsion, so that a gaseous mass has a continual tendency to expand to a greater and greater volume. This property, called the expansibility of gases, is commonly illustrated by the following experiment:—

A bladder, nearly empty of air, and tied at the neck, is placed under the receiver of an air-pump. At first the air which it contains and the external air oppose each other by their mutual pressure,

Fig. 11.—Expansibility of Gases.

and are in equilibrium. But if we proceed to exhaust the receiver, and thus diminish the external pressure, the bladder gradually becomes inflated, and thus manifests the tendency of the gas which it contains to occupy a greater volume.

It follows from this property that, however large a vessel may be, it can always be filled by *any quantity whatever* of a gas, which will always exert pressure against the sides. It is in consequence of the existence of this pressure, which is itself a result of expansibility, that the name of *elastic fluids* is often given to gases.

It is necessary to remark that the same substance may, according to its temperature, assume any one of the three states. Thus water in the cold of winter assumes the solid state and becomes ice; and, on the other hand, there is always more or less water diffused through the air in the gaseous state, called aqueous vapour. If the thermal conditions existing at the surface of the earth were to receive a notable change in either direction, some of the bodies which we habitually see in the liquid state would become either solids or vapours.

19. Molecular Constitution.—Whatever be the state under which a body presents itself, it is the general opinion of physicists that it is not composed of continuous matter, but is an aggregation of distinct parts held at a distance from each other. These constituent parts are called particles or molecules. They must be regarded as exercising two kinds of mutual actions, the one attractive, the other repulsive, which balance each other in the case of solids and liquids. In the case of gases this equilibrium does not subsist; there is a permanent repulsive force between the particles, which gives rise to expansibility or elastic force.

The molecules[1] of solids and of liquids ought not to be considered as similar. In the latter, in fact, each molecule can turn on its axis without producing any modification in the equilibrium; in other words, equilibrium depends only on the molecular distances and not at all on the form or relative disposition of the molecules. An approximate idea of this physical constitution will be obtained by assuming that the molecules of liquids are spherical, and hence that molecular equilibrium depends only on the distances between the centres of the spheres.

In solids, much depends upon the form and relative disposition of the molecules. It would seem as if these molecules (according to the ideas of some ancient philosophers) were formed with hooked projections which become locked together and so give a determinate figure to the mass. It is not, however, necessary to fall back upon such a gross image as this for the explanation of rigidity. It is sufficient to conceive that when an effort is exerted against any part of a solid body, its molecules turn on their axes, assume new directions, and take up a new position of equilibrium. Such a supposition corresponds with that invariability of form which we are accustomed to connect with the solid state. In reality this invariability is not absolute. The smallest force applied to a solid body produces some change of form, but frequently this change is only appreciable when the force is very intense.

20. Divisibility.—This hypothesis regarding the constitution of bodies amounts to assuming that matter is not infinitely divisible, but that, whatever be the means employed to produce division, there is for each body a limit below which it never descends. These

---

[1] The hypotheses broached in the remainder of this section must be received with caution, as being merely conjectural explanations of the distinction between solids and liquids.

parts which always remain undivided are called atoms (α, privative, τέμνω, to cut); and we are to understand by this designation not elements which it is *impossible* to divide in an absolute or metaphysical sense, but elements which are not susceptible of division by any known forces.

It is in chemistry especially that reasons are found for assuming a limit to the division of matter. In fact in chemical phenomena we see this division attain limits which, though doubtless very remote, are yet fixed. The composition of compound bodies is invariable, whatever may be the circumstances of their production;[1] their properties, which could hardly fail to be altered by a change in the size of the constituent particles, are also the same; whence it seems necessary to conclude that the elements between which chemical affinity is exerted are absolutely alike and unchangeable—have a definite existence. They are the veritable *individuals* of the mineral kingdom. These are what we mean by atoms.

The notion of an atom does not involve the idea of size; but experience teaches us that their size must be excessively minute; for we can in several different ways divide matter into extremely small parts, without finding any reason to think that we have attained or even approached the limit of division. We will cite some examples which prove the extreme divisibility of matter.

Wollaston succeeded in obtaining threads of platinum of a diameter not exceeding $\frac{1}{30000000}$ of an inch. The method which he employed for preparing them consisted in drawing a silver wire with a platinum core, and dissolving the shell of silver in nitric acid. In this way threads can be obtained so fine that they are actually invisible to direct view, and that their existence can only be detected by the aid of certain special optical phenomena. In the art of beating gold, leaves are obtained whose thickness cannot exceed $\frac{1}{200000}$ of an inch. A square inch of this leaf would weigh less than the $\frac{1}{30000}$ of an ounce, and as a square whose side is $\frac{1}{250}$ of an inch is visible to the naked eye, it follows that this square inch of leaf contains more than 60,000 visible parts.

The diffusion of colouring matters and perfumes affords a notable instance of the extreme divisibility of matter. A cubic millimetre of indigo (about $\frac{1}{16000}$ of a cubic inch) dissolved in sulphuric acid,

---

[1] This is called the law of "definite proportions." The laws of "multiple proportions" and "equivalent proportions" furnish perhaps a still stronger argument for the atomic hypothesis.

can colour to an appreciable extent more than 10 litres (about 2 gallons) of water. Now, a litre contains a million cubic millimetres; the cubic millimetre of indigo, therefore, in this experiment is divided into ten million visible parts.

The diffusion of odoriferous substances is still more astonishing. It is well known that a grain of musk will continue for years to supply the air, which is continually being renewed around it, with a sufficient number of particles to communicate to it its odour. The mind can hardly form an idea of the degree of tenuity which such particles must have.

21. Porosity.—Porosity is an immediate consequence of the hypothesis of molecular constitution. It consists in the existence, in the interior of all bodies, of intervals or pores between their material particles. This porosity is often so marked as to permit the passage of liquids or gases through the substance of solids. It then receives the name of *permeability*. Permeability is the property which is utilized in the employment of stone filters; the pores are large enough to allow the water to pass, and small enough to prevent the passage of the small solid bodies held in suspension in the water.

It is by means of permeability that the communication and contact of liquids takes place in organized bodies; for the vessels which contain them are nowhere open, and it is always through the substance of their walls that the final changes of elements are made which are necessary to vital action.

By using great pressure, liquids can be made to pass through metals. These latter, or at least some of them, iron and platinum for example, when raised to a high temperature, allow ready passage to different gases. Thus it is that cast-iron stoves, when red-hot, allow some of the deleterious products of combustion to pass out, and sometimes occasion serious accidents.

But even when no permeability can be detected, pores must still be assumed to exist; and the proof is found in the fact that all bodies can have their volumes increased or diminished,—that they are dilatable and compressible. The dilatation of bodies by the action of heat is a general phenomenon which will be studied further on. We will here confine ourselves to compressibility.

22. Compressibility.—Compressibility consists in the reduction of volume which bodies experience under the action of external pressure. The compressibility of solids is extremely small; that is to say, a very considerable pressure is required to produce any sensible diminution of volume. The existence of such an effect has, however, been well established, and it is found necessary to allow for it in structures.

The compressibility of liquids, though greater than that of solids, is still very small. Hence, in comparison with gases, they have often been called incompressible fluids. It is easy with the aid of

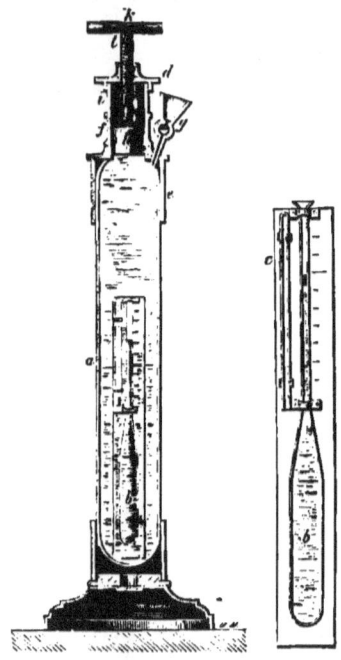

Œrsted's apparatus, represented in section in Fig. 12, to show that liquids are more compressible than solids, and to measure approximately the degree of their compressibility. The liquid to be compressed is contained in a kind of large thermometer b, called a piezometer, whose tube has been carefully divided, and its reservoir gauged so as to determine how many divisions of the tube its volume is equal to. The tube is open at the top, and a globule of mercury, placed above the liquid column, serves for index. The apparatus is placed in a vessel of water a, having very thick sides.

When pressure is exerted by means of the screw-piston klh, the index of mercury is seen to descend, showing a diminution of the volume of the liquid. The

Fig. 12.—Œrsted's Piezometer.

amount of the pressure is known from the volume of air contained in the tube c, which serves as a manometer. In this experiment the number of divisions through which the extremity of the liquid column moves indicates the *apparent* diminution of volume; that is to say, the excess of the diminution of volume of the liquid above that of the envelope. It is easy, in fact, to understand that the piezometer itself must, under the pressure to which it is subjected, undergo a diminution of capacity, which must be taken into account. Œrsted supposed that this diminution was insensible, or nothing, since the pressure is exerted on the interior as well as the exterior of the piezometer. But this conclusion is erroneous; for if the piezometer were solid and submitted to compression, the interior shells would react with a force precisely equivalent to that which is produced when the instrument is hollow and the liquid occupies its interior. The piezometer then, in Œrsted's experiment, undergoes a diminution of volume equal to that which a solid piezometer would undergo in the same circumstances. In order, then, to find the true diminution of volume of the liquid, it is necessary to increase the apparent contraction by the contraction of the envelope. This latter element varies according to the quality of the glass of which the envelope is composed, but may be estimated at about ·0000029 per atmosphere of pressure. The true compressibility of water, according to recent experiments conducted under the direction of M. Jamin by Messrs. Amaury and Descamps, is, at the temperature of 15° centigrade, ·0000457 per atmosphere. It diminishes when the temperature increases. Alcohol and ether are rather more compressible, and their compressibility (unlike that of water) increases with the temperature. Mercury is much less compressible than water. Its variations of volume may therefore, in ordinary experiments, be neglected.

Gases are enormously more compressible than solids and liquids. This is easily shown by the pneumatic syringe, Fig. 13. It is a cylinder of very thick glass, closed at one end. A piston, which exactly fits the tube, is made to enter the other end, and can be forced in until the air is reduced to a half, a third, or a tenth of its original volume.

Hence it would seem to follow that in gases the spaces between the particles are much greater than in liquids and solids, and consequently that there is much less matter in the same volume. The same conclusion is established by the comparison of specific gravities. To give an idea of the difference, it may suffice to mention that water, when converted into steam, at the ordinary atmospheric pressure, and at the ordinary temperature of boiling water, expands 1700 times, so that a cubic inch of water gives about a cubic foot of steam.

Fig. 13.—Pneumatic Syringe.

23. Elasticity.—This term, when applied to solids, is used in modern physics to denote the property in virtue of which a body tends to recover its form and dimensions when these are forcibly changed. The great majority of solid bodies possess almost perfect

elasticity for small deformations; that is to say, when distorted, extended, or compressed within certain small limits, they will, on the removal of the constraint to which they have been subjected, return instantly to their original form and dimensions. These limits (which are called the limits of elasticity) are different for different substances; and when a body is distorted to an extent exceeding these limits, it takes a set, the form to which it returns being intermediate between its original form and that into which it was distorted.

When a body is distorted (using this word to include extension or compression as well as change of shape) within the limits of its elasticity, the force with which it reacts is simply proportional to the amount of distortion. For example, the force required to make the prongs of a tuning-fork approach each other by a tenth of an inch, is precisely double of that required to produce an approach of a twentieth of an inch; and if a chain is lengthened a twentieth of an inch by a weight of 1 cwt., it will be lengthened $\frac{1}{10}$th of an inch by a weight of 2 cwt., the chain being supposed to be sufficiently elastic to experience no permanent set from this greater weight. Also (within the limits of elasticity) equal and opposite distortions are resisted by equal reactions; for example, the same force that suffices to make the prongs of a tuning-fork approach by $\frac{1}{10}$th of an inch, will suffice, if applied in the opposite direction, to make them separate by the same amount.

An important consequence which can be mathematically deduced from the laws just stated, is that when a body is distorted within its limits of elasticity, the vibrations which ensue when the constraint is removed have periods which are independent of the magnitude of the distortion. For example, a common C tuning-fork makes about 528 vibrations in a second whether vibrating strongly or feebly; by whatever amount the prongs are made to approach each other, the time which elapses from their being released to the attainment of their greatest separation is $\frac{1}{1056}$ of a second, and the same time elapses from their greatest separation to their nearest approach. The sum of these two intervals, $\frac{1}{528}$ of a second, is the period of a complete vibration; and during the whole time that the vibrations are dying away until the fork finally comes to rest, this period remains unaltered, the diminution of distance moved being exactly compensated by the increasing slowness of the motion.

India-rubber has very wide limits of elasticity. Glass has sensibly perfect elasticity up to the limit at which it breaks.

Putty and wet clay are instances of bodies which are almost entirely destitute of elasticity.

The resistance of a cylindrical or prismatic bar to elongation or flexure is measured by a number called "Young's modulus of elasticity." The following are examples of its value for different substances, in kilogrammes per square millimetre:—

Flint-glass,......................5,851	Iron (wrought),..............19,994
Brass,........................10,943	Do. (cast),..................13,741
Steel,........................21,793	Copper, ...12,558

To illustrate the meaning of this table by the case of steel; a steel wire whose section is a square millimetre will be elongated by $\frac{1}{21793}$ of its length by a weight of one kilogramme. The elongation is inversely proportional to the section and directly proportional to the stretching weight; so that a steel wire whose section is half a millimetre will, when stretched by a weight of six kilogrammes, receive twelve times the elongation above specified.

The resistance of a cylindrical or prismatic bar or beam to bending (called its flexural rigidity) is proportional to the value of Young's modulus of elasticity for the material of the bar or beam; so that from the dimensions of the bar, the value of this modulus, and the magnitudes and directions of the externally applied forces, the amount of bending could be calculated.

The resistance of a cylindrical rod to twisting (called torsional rigidity) does not depend upon the value of Young's modulus, but upon an entirely distinct element, an element which is sometimes called simply "rigidity," and which expresses the resistance which a square of given thickness would oppose to being changed by external forces into a rhombus of the same area, having angles differing by a given small amount from right angles.

Elasticity being a molecular phenomenon, it is to be expected that all circumstances which modify the molecular constitution of a body will alter its elasticity; but in the present state of science it is impossible to predict à priori the nature and direction of the change, the effects being sometimes opposite for different substances. Thus tempering (that is to say heating followed by sudden cooling), which, as is well known, augments in a high degree the hardness and elasticity of steel, produces a reverse effect on the bronze of which gongs are made. This alloy, in fact, when cooled slowly, possesses the fragility of glass; whilst, when cooled suddenly, it can be wrought with the hammer.

The elasticity of springs furnishes a simple means of comparing forces. Fig. 14 represents an apparatus designed for this purpose and called a *dynamometer*. It is formed of two plates of steel, AB and A'B', jointed at their extremities to two metallic bridles which connect them. To the middle of the upper plate is attached a ring, by means of which the apparatus can be suspended from a fixed

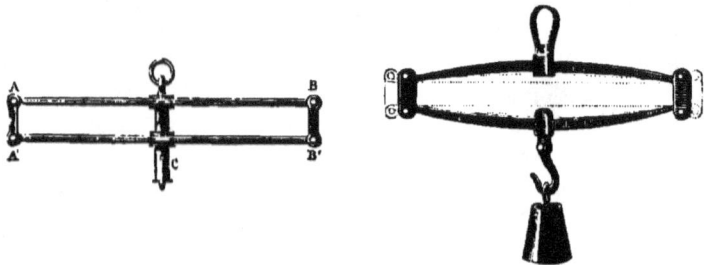

Fig. 14.—Dynamometer.

point. To the middle of the lower plate is attached a hook, which can either receive a weight or serve as a point of application for the force which is to be tested. Under the action of the force thus applied the spring plates bend, the distance of the middle points increases, and this increase serves to measure the force itself; or the force may be measured by observing what weight must be suspended from the hook to produce the same effect.

The spring-balance is another apparatus of the same kind. In its most common form it contains a spiral spring which is elongated by the application of the force which is to be measured. The equality of the graduations illustrates the law above stated of the proportionality of distortions to the forces producing them. The resistance of a spiral spring to elongation depends chiefly (as shown by Professor James Thomson) on the torsional rigidity of the wire which composes it.

It is important to remark, that whereas a pair of scales is essentially a measure of mass, a spring-balance is essentially a measure of force. Hence if a spring-balance be graduated so as to show weights correctly at a medium latitude, it will indicate too little if carried to the equator (where the force of gravity is feebler), and too much at the poles (where gravity is more intense).

CHAPTER IV.

GRAVITY.

24. Terrestrial gravity is the force in virtue of which all bodies fall to the surface of the earth. This force is general; its effects are observed in all places and for all bodies. If some of these latter, as smoke and hydrogen gas, appear to be exceptions, it is because they are sustained by the air in the same manner as cork is sustained by water. In space deprived of air, not only do all bodies fall, but, as we shall see later, they fall with equal velocities.

25. Direction of Gravity.—The direction of gravity is called the *vertical*. It is easily determined by the aid of the simple apparatus called a plumb-line, which consists of a thread fixed at one end and carrying a heavy body at the other. When the system is in equilibrium it is clear that the resultant of the actions of gravity on all the parts of the heavy body has exactly the same direction as the thread, since it is this which prevents the fall. But it can be shown that this direction does not change when the form and volume of the heavy body are altered, it must therefore be the same as the direction of the force which would act upon one of the elementary particles if suspended alone at the extremity of the thread.

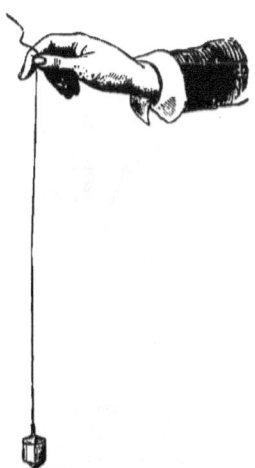

Fig. 15.—Plumb-line.

It can be shown by experiment that the direction of gravity is perpendicular to the surface of a liquid in equilibrium, or to use the common expression, to the surface of still water. For this purpose a plumb-line OA is suspended over the

surface of a fluid in equilibrium (which should be slightly opaque,
as blackened water), and the
plummet is allowed to plunge
in the liquid. The image AB
of the thread produced by re-
flection at the surface of the
liquid will be seen with
great distinctness, and will
be observed to be exactly in
a line with the thread itself.
Now we shall see in a subsequent part of this treatise, that whenever reflec-
tion takes place at a plane
mirror, each point of the object and the corresponding
point in the image are on
the same perpendicular to
the mirror and at equal distances from the mirror on opposite sides. Since, then, in
the experiment here described
the thread and its image are

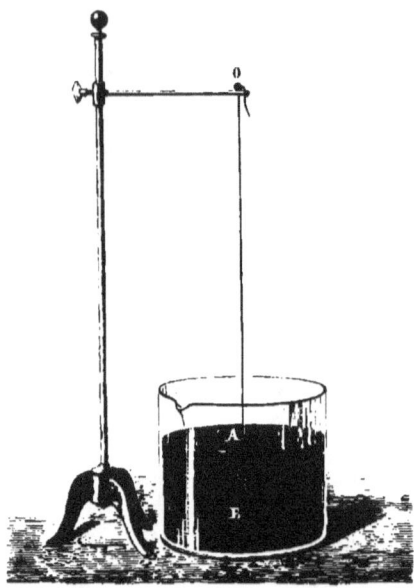

Fig. 16.—Experiment for showing that the Plumb-line is
perpendicular to the surface of a fluid at rest.

in one straight line, this line must be perpendicular to the surface.

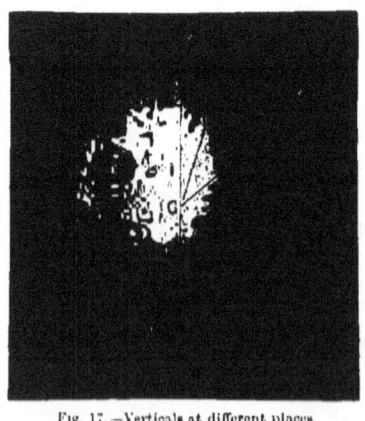

Fig. 17.—Verticals at different places.

The surface of still water de-
fines in each locality what is
called the surface of the earth.
This expression denotes the surface of an imaginary ocean of
calm water supposed to cover
the whole earth. This surface
is known to be sensibly spherical. It follows that the different verticals will nearly meet
in the centre of the earth. The
figure shows the relative position of some verticals CZ, CZ',
CZ''; it is evident that they
contain angles equal to the an-
gular distance which separates the corresponding places.

At any one locality all verticals may be treated as parallel, on account of the immense distance of the centre of the earth. Let us calculate, for example, the angle contained between two verticals a metre (39·37 inches) apart. Ten millions of metres correspond to a quarter of the earth's circumference, that is to say, to 90°. A length of a metre, therefore, represents 90° divided by ten millions, that is to say, about $\frac{3}{100}$ of a second, a quantity quite inappreciable even with our most perfect instruments. It should be remarked, however, that the parallelism of the verticals at any one place is a physical fact, completely independent of all previous knowledge of the figure of the earth, and can be established by direct observation.

We may remark in passing, that the latitudes of places on the earth's surface are determined by the directions of the verticals. What is commonly called the latitude of a place is the angle which a vertical at the place makes with a plane perpendicular to the earth's axis of rotation. As distinguished from this, the geocentric latitude (which is required in a few astronomical problems) is the angle which a line drawn from the place to the earth's centre makes with a plane perpendicular to the axis of rotation. The difference between common and geocentric latitude generally amounts to some minutes, and attains its greatest value (11′ 29″) at latitude 45°.

26. Point of Application of Gravity—Centre of Gravity.—Gravity being a property of matter, its points of application must evidently be the different material particles which compose each body. Though a body be divided into as many parts as we please, and even reduced to the state of impalpable powder, each of the grains thus obtained will be subject to the action of gravity. The total force which urges a body to fall is the sum, or more strictly, the resultant of all the forces which are thus actually applied to its several elements. Now these forces are parallel, as has just been stated, and act in the same direction; their resultant is therefore equal to

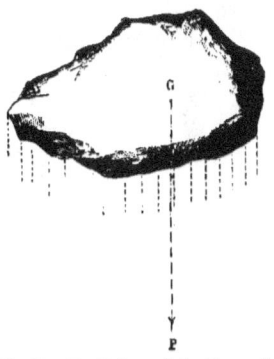

Fig. 18.—Parallelism of the Forces of Gravity on the different Points of a Body.

their sum, and it constitutes what is called the *weight* of the body; that is to say, the force with which it presses the obstacle which prevents it from falling. The point of application G of this resultant

3

(Fig. 18) is called the *centre of gravity*. It follows from the property indicated in section 15, that the position of this point does not vary when the direction of the components is made to vary. The body can therefore be turned about in any manner without the centre of gravity changing its position in the body. It is a fixed point depending only on the form of the body and the distribution of the matter which composes it.

If the body has the same density throughout, the position of the centre of gravity depends only on the figure, so that in this case bodies of similar form have their centres of gravity similarly situated.

The determination of the position of the centre of gravity is a problem of mechanics which is solved by appropriate methods of general application, founded on the principle (to which we can here barely allude), that if a body be divided into a great number of equal elements, the sum of their distances from any one plane, divided by the number of elements, is equal to the distance of the centre of gravity from the same plane.

Whenever a body of uniform density contains a point which is a centre of symmetry (that is, a point which bisects all straight lines drawn through it), this point must be the centre of gravity. Hence:

1. The centre of gravity of a straight line is its middle point;

2. The centre of gravity of a circle, or of the circumference of a circle, is the centre;

3. The centre of gravity of a parallelogram is the intersection of the diagonals;

4. The centre of gravity of a sphere is its centre;

5. The centre of gravity of a cylinder is the middle point of its axis;

6. The centre of gravity of a parallelopiped is the common intersection of the diagonals, &c.

It may naturally be asked how we can speak of centres of gravity of lines and surfaces, which, being only of one or two dimensions, cannot possess weight. The answer is, that in so speaking we make an abstraction analogous to that which gives us the idea of a material point. We suppose lines or surfaces composed of elements possessing weight; and the results thus obtained can be utilized in investigating the centres of gravity of real bodies. Consider, for example, a triangular prism. It can be conceived as decomposed into elements which would be, so to speak, heavy triangles. The centre of gravity of the solid would therefore be in the line which joins the centres of gravity of all these triangles, and would be its middle point. We may say then that the centre of gravity of a triangular prism, and in general of any prism, if composed of homogeneous material, is the middle point of the line which joins the centres of gravity of the two ends.

We must guard against the error of supposing that the force of gravity on a body acts at the centre of gravity. Gravity really acts equally on all the particles of the body; but its effect is in many respects the same as that of a single imaginary force supposed to act at the centre of gravity.

The centre of gravity sometimes lies outside the body, as in the case of a ring or a hollow sphere. When this is the case, it must be regarded as rigidly connected to the body.

27. Physical Definition of the Centre of Gravity.—The centre of gravity, regarded from the mechanical point of view, is in reality merely the centre of parallel forces distributed in a determinate manner. Its position can therefore be found by a purely geometrical investigation, and apart from any physical idea of the nature of bodies. Nevertheless, it is certain that the discovery of the existence of centres of gravity had its origin in the consideration of the phenomena of equilibrium, which are exhibited by bodies under the influence of gravity. Experiment, in fact, shows that in most bodies there is a point such that, if it be supported, the body will be in equilibrium, and if it be not supported, the body will move under the action of gravity. This point is the centre of gravity, and the property in question is an obvious consequence of the mechanical definition already given. But this property may itself be used as the definition of the centre of gravity (though with some want of precision). We shall thus use it in the examination of some important cases of equilibrium.

28. Equilibrium of a Body capable of turning about an Axis or a Fixed Point.—Consider, for example, a triangular plate movable about an axis of rotation O, and let G be the position of the centre of gravity. In order that there may be equilibrium, the centre of gravity must be supported; that is to say, the vertical drawn through it must meet the axis. This condition may be fulfilled by two very different positions of the body: the centre of gravity may be either above or below the axis. In the former case (Fig. 20) it is evident that if the body be ever so little displaced from its position of equilibrium, the effect of gravity will be to make it fall still further away. In the second case, on the contrary (Fig. 19), the action of

gravity tends to restore the body to the position of equilibrium.   In
the former case the equilibrium is *unstable;* in the latter it is *stable.*

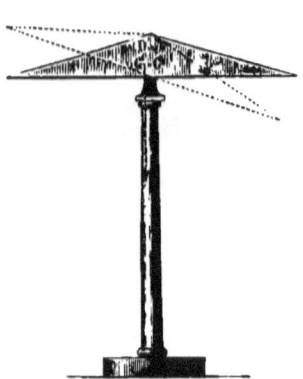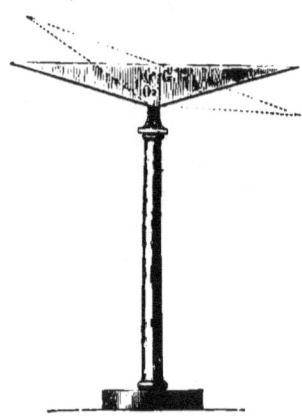

Fig. 19.—Stable Equilibrium.   Fig. 20.—Unstable Equilibrium.

We see then that the condition of stable equilibrium is that the
centre of gravity be below the axis
or point of suspension.

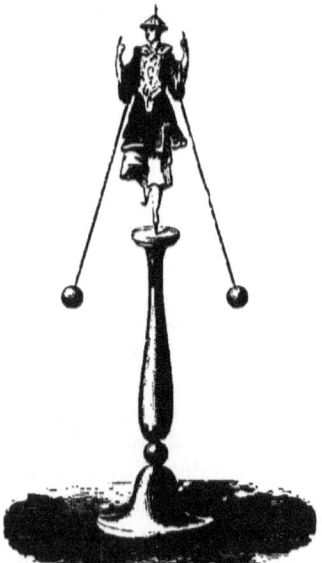

Fig. 21.—Balancer.

The toy called the balancer is an
application of this principle.   It
consists of an ivory figure resting
by one point on a small horizontal
stand.   Two stiff wires fixed to the
figure terminate below in leaden
balls.   The centre of gravity of the
system is thus brought below the
point of support; the equilibrium
is consequently stable.   If we draw
the figure to one side and then re-
lease it, it will perform a series of
oscillations, and will end by taking
a position of equilibrium such that
the vertical through the centre of
gravity passes through the point of
support.

If a body were traversed by an
axis through its centre of gravity,
its equilibrium would be neutral, and the body would remain in

equilibrium in all positions. This condition ought to be rigorously fulfilled by the wheels of pieces of mechanism which only serve to transmit motion.

29. **Equilibrium of a Body resting on a Horizontal Plane which touches it in one Point.**—Consider (Figs. 22 and 23) a body of

Fig. 22.—Unstable Equilibrium.   Fig. 23.—Stable Equilibrium.

ellipsoidal form resting on a horizontal plane. In order that there may be equilibrium it is evidently necessary and sufficient that the vertical through the centre of gravity G should meet the horizontal plane at the point of contact. We see by the figure that this condition may be realized in two ways.

The first figure corresponds to *unstable*, the second to *stable* equilibrium. The figure shows that in the latter case the *centre of gravity occupies its lowest possible position.*

Fig. 24.—Tumblers

The tumbler (Fig. 24) is founded on this principle. The centre of gravity being near the lower side in consequence of the accumulation of matter in this region, the apparatus is in stable equilibrium. If it be removed from its position and subjected even to very wide displacements, it always rises again and returns to its position of equilibrium after a number of oscillations.

If the centre of gravity were always at the same distance from the plane—if, for example, the body were spherical, there would be equilibrium in all positions,—the equilibrium would be neutral.

We may remark in general that a position of unstable equilibrium is only mathematically possible; it never can have a physical existence; for the smallest derangement destroys it, and in nature a multitude of causes, such as the movement of the air, the flexibility of supports, &c., introduce displacements which violate the conditions of equilibrium. If there be any actual cases of such equilibrium— if, for example, it is possible to make an egg stand on its end, it is by the help of friction, which constitutes a new force tending to prevent displacement.

30. Equilibrium of a Body resting on a Horizontal Plane at several Points.—When a body rests on a horizontal plane at several points,

Fig. 25.—Equilibrium of a Body supported on a Horizontal Plane at three or more Points.

it is necessary for equilibrium that the vertical through the centre of gravity fall within the convex polygon which can be formed by joining the points of support. It is clear that in this case gravity will have no effect but to press the body against the plane. It is also obvious that the equilibrium will be the more stable as the centre of gravity is lower, and the distance of the vertical through it from the nearest side of the polygon greater. If this vertical is very near one side, a small force will be sufficient to overturn the body on that side, although a very great force may be required to overturn it in the opposite direction.

31. Practical Method of finding the Centre of Gravity.—The different methods which are employed in practice for the experimental determination of the centre of gravity are dependent on the principles above explained. Whatever be the particular nature of the proceeding, it always consists in placing the body in a position of equilibrium from which it can be inferred that the centre of gravity lies in a certain line or surface.

Thus, for example, if we suspend a body by one point, it is clear that the centre of gravity must lie in the prolongation of the suspending thread. If we then suspend the body by another point, a similar inference follows. Consequently, the centre of gravity must be the intersection G of the two directions thus indicated.

If we wish, for example, to pierce a plate or board by an axis which is to pass through its centre of gravity, we may begin by balancing it in a horizontal position upon two points near its circumference. The line joining them will pass vertically under the centre of gravity. By repeating the operation we may find a

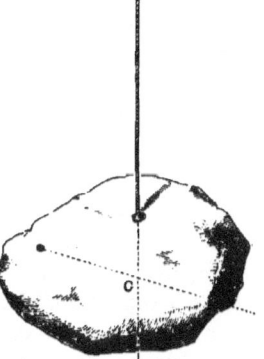

Fig. 26.—Experimental Determination of Centre of Gravity.

second line which possesses the same property, and the required axis must pass through their intersection and be perpendicular to the plate. Instead of balancing the plate upon two points, an operation which may require repeated trials, it is more expeditious, when practicable, to suspend it freely in a vertical position by a point near its circumference, and to suspend a plumb-line from the same point. The course of this line must be marked on the plate, and the operation must then be repeated, using a different point of suspension. The intersection of the two lines thus obtained will, as before, be opposite to the centre of gravity of the plate. Both the methods described in this paragraph are applicable even to plates which are not homogeneous.

CHAPTER V.

32. In air, bodies fall with unequal velocities; a sovereign or a ball of lead falls rapidly, a piece of down or thin paper slowly. It was formerly thought that this difference was inherent in the nature of the materials; but it is easy to show that this is not the case, for if we compress a mass of down or a piece of paper by rolling it into a ball, and compare it with a piece of gold-leaf, we shall find that the latter body falls more slowly than the former. The inequality of the velocities which we observe is due to the resistance of the air, which increases with the extent of surface exposed by the body.

It was Galileo who first discovered the cause of the unequal rapidity of fall of different bodies. To put the matter to the test, he prepared small balls of different substances, and let them fall at the same time from the top of the tower of Pisa; they struck the ground almost at the same instant. On changing their forms, so as to give them very different extents of surface, he observed that they fell with very unequal velocities. He was thus led to the conclusion that gravity acts on all substances with the same intensity, and that in a vacuum all bodies would fall with the same velocity.

This last proposition could not be put to the test of experiment in the time of Galileo, the air-pump not having yet been invented. The experiment was performed by Newton, and is now commonly exhibited in courses of experimental physics. For this purpose a tube from a yard and a half to two yards long is used, which can be exhausted of air, and which contains bodies of various densities, such as grains of lead, pieces of paper, and feathers. When the tube is full of air and is inverted, these different bodies are seen to fall with very unequal velocities; but if the experiment is repeated after the tube has been exhausted of air, no difference can be perceived between the times of their descent.

33. Laws of Falling Bodies.—Having found that the effect of gravity is the same on all bodies, Galileo proposed to himself the problem of determining, by experiments on one body, the law which regulates their descent; and, inasmuch as the observation of a body falling freely is very difficult, on account of the rapidity of its motion, he adopted a method of diminishing this rapidity without in other respects altering the law of motion. This method consisted in the use of the inclined plane.

Consider, in fact, a heavy body M, free to move along the inclined plane ABC. The weight of the body M being represented by MP, it can, (by § 16), be decomposed into two other forces, viz. MN perpendicular to the plane, which is destroyed by the resistance of the plane itself, and MT parallel to the plane, which alone produces the motion. Now this latter force is less than MP, but is a constant fraction of it, for at all points in the plane the parallelogram of forces will have the same form, and the ratio of MT to MP will be constant. This ratio is in fact the same as that of the height AC of the plane to its length AB, or in other words is the sine of the inclination of the plane to the horizon. The motion will therefore be less rapid, but will follow the same law as that of a body falling freely, and will be much easier to observe. The diminution of velocity has the further advantage of diminishing the relative importance of the resistance of the air, which increases very rapidly with every augmentation of velocity.

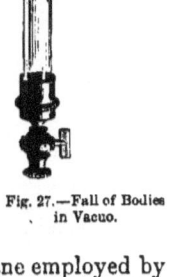

Fig. 27.—Fall of Bodies in Vacuo.

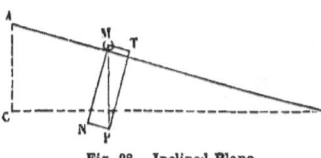

Fig. 23.—Inclined Plane.

The inclined plane employed by Galileo consisted of a long ruler, with a longitudinal groove, along which he caused a small heavy ball to roll. Having thus observed the spaces traversed in 1, 2, and 3 units of time, he found that these spaces were in the ratio of the numbers 1, 4, and 9; that is to say,

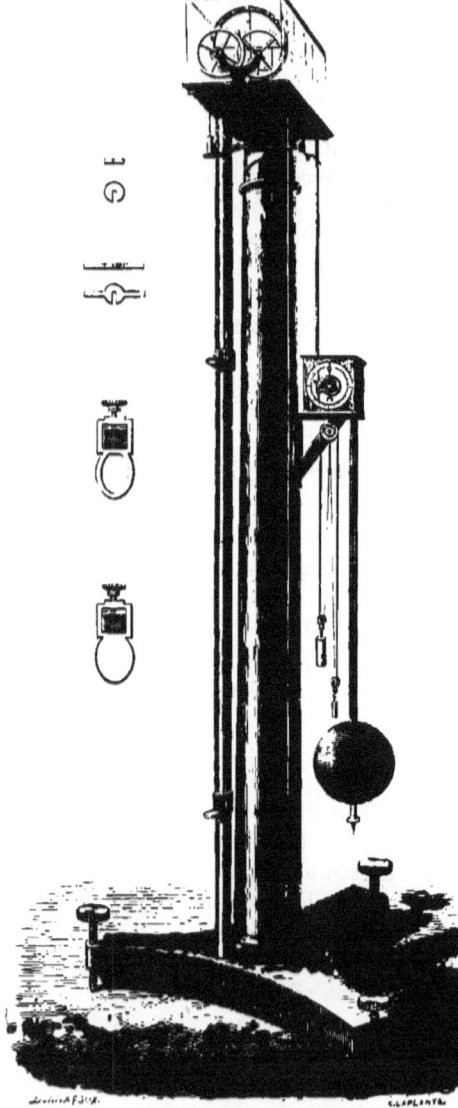

when the time of descent was doubled or tripled the space traversed became 4 or 9 times greater. This law can be expressed by saying that *the spaces traversed are proportional to the squares of the times of descent.*

34. Attwood's Machine. — Attwood, a fellow and tutor of Trinity College, Cambridge, invented, towards the end of last century, a machine which affords great facilities for verifying the laws of falling bodies. It involves, like Galileo's inclined plane, a method of diminishing the velocity of descent; but this result is obtained by very different means.

The machine consists of a column, having at its top a very freely moving pulley, which forms the essential part of the apparatus. In order to obtain great freedom for the movements of the pulley, the ends of its axis are made to rest, not on fixed supports, but on the

Fig. 29.—Attwood's Machine.

circumferences of wheels (two at each end of the axis) called friction-wheels, because this arrangement produces a great diminution of friction. Over the pulley passes a fine thread, carrying at its extremities two equal weights P. Neglecting the weight of the thread, it is obvious that these weights will be in equilibrium in every position. If however one of them be loaded with an additional weight p, the system will be put in motion, and all parts of it will move with the same velocity. We may therefore regard the moving force as distributed uniformly through it. But this force is simply the weight of p. If then, for example, the movable system $2P + p$ has 20 times the weight of p, each portion of the system is urged with a force equal to $\frac{1}{20}$ of its own weight. The force which produces motion is in general diminished, as compared with a body falling freely, in the ratio expressed by the fraction $\frac{p}{2P + p}$; and as this ratio continues constant through the whole motion, the law of the motion will be the same as that for free descent.

The following are the arrangements for observing the motion:—One of the weights moves in front of a graduated scale, and a plane stop for intercepting the descending weight can be fixed at pleasure at any part of this scale. A clock with a pendulum beating seconds serves for the measurement of time. To measure the space traversed in a second, the weight is raised to the commencement of the graduation, is then loaded with the additional weight, and is dropped precisely at one of the beats of the pendulum. The stop is placed by trial at such a point of the scale that the blow of the weight against it precisely coincides with another beat of the pendulum,—a coincidence which can be obtained with great accuracy, inasmuch as the ear easily detects the smallest interval between the two sounds. In order to insure a similar coincidence at the commencement of the fall, the weight is supported by a movable platform M (Fig. 30), which is prevented from falling by the upper end of the lever aob, whose lower end is guided by a cam[1] fixed to the escapement wheel

Fig. 30.—Detent in Attwood's Machine.

---

[1] A cam is a rotating piece which, by means of projections or indentations in its outline, guides the movements of another piece which presses against it.

of the clock, and is kept constantly pressed against the cam by means of a spring not shown in the figure. Suppose the wheel to be turning in the direction indicated by the arrow. It is obvious that as soon as the tooth of the cam has passed the end of the lever the latter will fly to the left, and therefore the upper end will fly to the right, since the lever turns about an axis at O. The platform M is thus suddenly dropped, and the fall commences. The position of the cam must be so adjusted that this movement shall take place exactly at the instant of the escapement of one of the teeth.

It is thus easy to measure the spaces traversed by the movable system of weights in 1, 2, and 3 seconds, and the result obtained will be as follows:—Suppose that in the first second the space traversed is 11 divisions, then we shall find:

$$
\begin{aligned}
\text{Space traversed in 2 seconds} &= 44 = 11 \times 2^2 \\
\text{Space traversed in 3 seconds} &= 99 = 11 \times 3^2 \\
\text{Space traversed in 4 seconds} &= 176 = 11 \times 4^2
\end{aligned}
$$

We see, then, that the spaces vary as the squares of the times employed in describing them. If we use the indefinite symbol K to denote the space described in the first unit of time, the space described in the time t will be given by the formula

$$s = Kt^2. \tag{1}$$

35. Velocities.—Attwood's machine also affords the means of studying the successive velocities which gravity imparts to the system. Before describing the means employed for attaining this end, it will be desirable to make a few remarks respecting velocity.

When a material point moves uniformly; that is to say, when it traverses equal spaces in equal times, the meaning of velocity is perfectly clear: it is the space traversed in unit time. Thus, if a point moving uniformly describes 2 feet in each second, we say that the velocity is 2 feet per second; or if it is understood that the foot and second are to be our units of space and time, we simply say that the velocity is 2. If a point moving uniformly describes 5 feet in 2 seconds, its velocity is $2\frac{1}{2}$, since the space described in one second must be half that described in two; and in general, in any case of uniform motion of a point, the velocity (in feet per second) will be obtained by dividing the whole space (in feet) by the whole time occupied in its description (in seconds).

But uniform motion is in nature the exception rather than the rule. In fact, it can only occur when the moving body is acted on either by no forces at all, or by forces in equilibrium. This, in fact, is merely a statement of the principle of inertia. When a body is constantly acted on by a force, this force must evidently have the effect of continually modifying the motion, and consequently the above mode of computing velocity is not directly applicable. If, however, we suppose the action of the force suddenly to cease, the motion will become uniform, and the velocity during this uniform motion will serve as a measure of the velocity which existed at the instant when the action of the force ceased.

Attwood's machine contains an arrangement for thus suddenly arresting the action of gravity at a given instant. For this purpose, a ring large enough to allow either of the two equal weights to pass through it, is to be fixed at the point of the scale at which the moving weight arrives at the end of one second. The additional weight, which for this purpose is made long and flat, is intercepted by the ring, and the subsequent motion, being due merely to the momentum already acquired, will be uniform. The stop is to be placed at the point at which the weight arrives a second later. The distance between the ring and the stop will then represent the velocity acquired during the first second. Making this experiment under the same conditions as the foregoing experiments on spaces, we find that the velocity acquired during the first second is represented by 22 divisions. We then place the ring at the point at which the system arrives after 2, 3, &c., seconds, and the stop at the point at which it arrives a second later; we thus measure the velocities acquired in 2, 3, &c., seconds, and find them equal to 44, 66, &c. We see, then, that the *velocities acquired in different times are proportional to the times.* Further, the velocity acquired in one second, 22 (divisions per second), is double of the space, 11 (divisions), described in the first second.

In formula (1), K denotes the space described in the first unit of time; the velocity acquired is then 2K, and consequently the velocity acquired in time t is given by the formula

$$V = 2Kt. \tag{2}$$

36. Attwood's machine, when fitted with the appliances above described, leaves little to be desired in point of accuracy, but its complication renders it expensive. We subjoin a figure representing Bourbouze's modification of Attwood's machine, which is much simpler. AB is the pulley, on the axis of which is a cylinder P,

surrounded with smoked paper. One of the iron weights, M, is held at the bottom of the apparatus by an electro-magnet, which is mag-

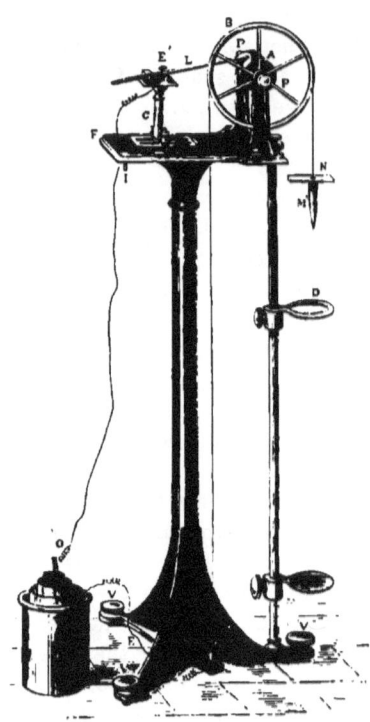

netized by means of the gal-vanic cell O. The weight M′, loaded with the additional weight N, is thus prevented from obeying the force of gravity. Again, the vibrating plate L, carrying a very light style for tracing a mark on the smoked cylinder, is held by the electro-magnet, E′, which is magnetized by the same cell. If at any moment the current is interrupted, the weight M′ falls, and the plate vibrates, describing an undu-lated curve on the surface of the cylinder. The undulations of this curve correspond to the vibrations of an elastic body, and, consequently, to equal times; while the distance of any undulation from the be-ginning of the curve is equal to the distance turned by the cylinder P, and is consequently proportional to the distance travelled by the weights.

Fig. 31.—Bourbouze's Modification of Attwood's Machine.

These distances are found to be exactly proportional to the series of numbers 1, 4, 9, &c. The ring D serves to intercept, at a given instant during the descent, the additional weight N; from this time onward the motion is uniform and the undulations of the curve are equidistant.

Attwood's machine, however modified, gives only indirect evidence regarding the motion of bodies falling freely. Although this cir-cumstance cannot affect the legitimacy or accuracy of the conclusions to which it leads, it would be interesting, if possible, to observe the phenomenon of free fall, and show that the laws just obtained are verified. This is the object of Morin's apparatus.

37. Morin's Apparatus.—Morin's apparatus consists of a wooden cylinder covered with paper, which can be set in uniform rotation about its axis by the fall of a heavy weight. The cord which supports the weight is wound upon a drum, furnished with a toothed wheel which works on one side with an endless screw on the axis of the cylinder, and on the other drives an axis carrying fans which serve to regulate the motion.

In front of the turning cylinder is a cylindro-conical weight of cast-iron carrying a pencil whose point presses against the paper, and having ears which slide on vertical threads, serving to guide it in its fall. By pressing a lever, the weight can be made to fall at a chosen moment. The proper time for this is when the motion of the cylinder has become sensibly uniform. It follows from this arrangement that during its vertical motion the pencil will meet in succession the different generating

Fig. 32.—Morin's Apparatus.

lines[1] of the revolving cylinder, and will consequently describe on its surface a certain curve, from the study of which we shall be

[1] That is, lines drawn on the surface of the cylinder parallel to its axis. A cylindric surface could obviously be described by the motion of a straight line in space. The line so moving is said to generate the cylindric surface, and the different positions which it successively occupies in its supposed motion are called generating lines of the cylinder. If a cylindric surface is cut open along a generating line and flattened out so as to become plane, its form will be rectangular, and its generating lines will be parallel to two sides of the rectangle.

able to gather the law of the fall of the body which has traced it. With this view, we describe (by turning the cylinder while the pencil is stationary) a circle passing through the commencement of the curve, and also draw a vertical line through this point. We cut the paper along this latter line and develop it (that is, flatten it out into a plane). It then presents the appearance shown in Fig. 33.

If we take on the horizontal line equal distances at 1, 2, 3, 4, 5 . . . ,

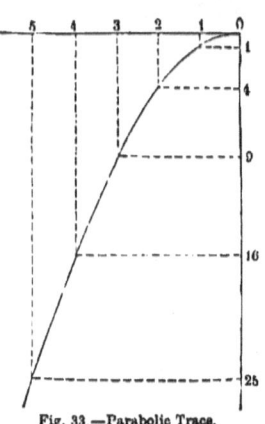

and draw perpendiculars at their extremities to meet the curve, it is evident that the points thus found are those which were traced by the pencil when the cylinder had turned through the distances 1, 2, 3, 4, 5. . . . The corresponding verticals represent the spaces traversed in the times 1, 2, 3, 4, 5. . . . Now we find, as the figure shows, that these spaces are represented by the numbers 1, 4, 9, 16, 25 . . . , thus verifying the principle that the spaces described are proportional to the squares of the times employed in their description.

Fig. 33.—Parabolic Trace.

We may remark that the proportionality of the vertical lines to the squares of the horizontal lines shows that the curve is a parabola. The parabolic trace is thus the consequence of the law of fall, and from the fact of the trace being parabolic we can infer the proportionality of the spaces to the squares of the times.

The law of velocities might also be verified separately by Morin's apparatus; we shall not describe the method which it would be necessary to employ, but shall content ourselves with remarking that the law of velocities is a logical consequence of the law of spaces.[1]

38. Formulæ relating to Falling Bodies.—The formulæ (1) and (2), §§ 34, 35, will enable us to obtain numerical solutions of questions relating to the fall of bodies, if we can ascertain the space traversed

---

[1] Consider, in fact, the space traversed in any time t, this space is given by the formula $s = Kt^2$; during the time $t + \theta$ the space traversed will be $K(t + \theta)^2 = Kt^2 + 2Kt\theta + K\theta^2$, whence it follows that the space traversed during the time θ after the time t is $2Kt\theta + K\theta^2$. The average velocity during this time θ is obtained by dividing the space by θ, and is $2Kt + K\theta$, which, by making θ very small, can be made to agree as accurately as we please with the value $2Kt$. This limiting value $2Kt$ must therefore be the velocity at the end of time t.—D.

by a falling body in one second, or (which will be numerically double of this) the velocity acquired in one second.

The several forms of apparatus which we have been describing furnish the means of making this determination, but not with much precision. A much better method of determination is furnished by the pendulum, and will be described in the next chapter.

The result obtained is that in Great Britain the velocity acquired by a body falling in vacuo for one second is 32·2 feet per second, or 9·81 metres per second. The velocity in question is usually denoted by the letter g, and is sometimes called the intensity of gravity, because its value for different localities varies in the same ratio as the force exerted by gravity on one and the same mass, and may therefore be taken as the numerical representative of the force of gravity on unit mass.[1]

The space traversed during the first second of fall is equal to $\frac{1}{2} g$, that is, to 16·1 feet.

Introducing g into the formulæ (1), (2), they become:

$$s = \tfrac{1}{2} \dot{g}\, t^2 \qquad\qquad (a)$$
$$v = g\, t \qquad\qquad (b)$$

Eliminating t from the equations (a), (b), we obtain a third formula, which gives the velocity acquired in falling through a given space:

$$v = \sqrt{2\,g\,s}. \qquad\qquad (c)$$

39. Applications.—I. To calculate the space traversed by a body which falls during a given number of seconds. This will be found from formula (a) by putting for t the given number of seconds. Performing the calculation as far as 10 seconds, we obtain the following table:—

TIME OF FALL, in Seconds.	SPACE FALLEN, in Feet.	TIME OF FALL, in Seconds.	SPACE FALLEN, in Feet.
1	16·1	6	579·6
2	64·4	7	788·9
3	144·9	8	1030·4
4	257·6	9	1304·1
5	402·5	10	1610·0

II. What is the time occupied by a body in falling from a height of of 750 feet, and what is the velocity with which it strikes the ground?

[1] If we adopt the "absolute" unit of force defined in § 42, the force of gravity on unit mass will be numerically equal to g.

4

Formulæ (a) and (c) give:

$$t = \sqrt{\frac{2s}{g}} = \sqrt{\frac{1500}{32\cdot2}} = 6\cdot8 \text{ seconds.}$$

$$v = \sqrt{2\,g\,s} = \sqrt{64\cdot4 \times 750} = 219\cdot8 \text{ feet per second.}$$

III. From what height must a body fall to acquire a velocity of 1500 feet per second?

Formula (c) gives:

$$s = \frac{v^2}{2\,g} = \frac{2250000}{64\cdot4} = 34938 \text{ feet.}$$

The velocity of 1500 feet per second is about that of a cannon-ball on leaving the muzzle of the gun. We see that it would be obtained by falling from a height of about 6½ miles. The duration of the fall would be about 47 seconds. It must be borne in mind that the formulæ (a), (b), (c) are only strictly applicable to fall in vacuo. In air they furnish results more and more remote from the truth, as the velocity increases.

In vacuo, a body thrown upwards from the earth would, on its return, strike the ground with a velocity equal to that with which it was thrown, and the velocity at any given point which it traverses both in its upward and downward course, would be the same in the descent as in the ascent. We see then from above that a cannon-ball would ascend to a height of about 6½ miles. In air, the velocity is slower in the descent than in the ascent, both because the height attained by the projectile is less than it would be in vacuo, and because the velocity acquired in falling from this diminished height is still further diminished by friction in the descent.

One notable difference between fall in vacuo and in air is, that in the latter case the velocity, instead of increasing indefinitely, only increases towards a certain limit which it can never exceed; and if a body be projected downwards with a velocity greater than this, its motion will be retarded. The resistance of the air, in fact, increases with the velocity, and the limit in question is that velocity at which the resistance encountered from the air is exactly equal to the weight of the body. The limiting velocity is not the same for all bodies, but depends on their sizes, densities, and forms.

39A. Motion of Projectiles.—When a body is projected in any direction, its subsequent motion (neglecting the resistance of the air) can be determined by means of the following principles:—

1. The horizontal component of motion will remain unchanged.

2. The vertical component of motion will be the sum or difference (according as the body was projected below or above the horizontal direction) of the motion of a body falling freely, and the vertical component of the initial motion.

These principles are sufficient to determine the position, the velocity, and the direction of motion of the body, after the lapse of any given interval, until it strikes an obstacle.

By reference to Fig. 33, it will easily be understood that, if the direction of projection be horizontal, the curve described will be a parabola; for, in virtue of the first principle, the body describes equal horizontal distances in equal times; and in virtue of the second principle (since the vertical component of initial motion is in this case zero), the vertical spaces described are those of a body falling freely; hence the construction of Fig. 33 is precisely applicable to this case.

If the direction of projection be oblique (that is, neither horizontal nor vertical), the path will still be parabolic. For example, if the body be projected obliquely upwards, we may divide its path into two parts, one described in its ascent, the other in its descent. These two parts will be precisely similar, and at the highest point of the path, where they join, the motion is horizontal. We may regard the curve in Fig. 33 as representing one of the two parts, say the descending part; for the motion, after passing the highest point, must evidently be the same as if the body had been projected horizontally from the highest point with the velocity which it actually has at that point. This velocity is, in fact, the horizontal component of the actual velocity of projection.

If a body be projected vertically downwards, its motion will be the same as if it had fallen from a certain height above. If it be projected vertically upwards, the times of ascent and descent will be equal, and the velocity at any one point will be the same in the descent as in the ascent. At the highest point the body is for an instant stationary; its descent is therefore the motion of a body falling freely. Formulæ (a), (b), (c) of § 38 will apply to the ascent as well as the descent, if in the former case we understand that the time denoted by t is the time reckoned backwards from the instant of attaining the highest point.

Whether the motion be in a vertical line, or in a parabola, the following principle will be found to apply, being in fact the mathematical consequence of the two principles already enunciated, viz. :—

The velocity is the same in the ascent and descent at any two points which are on the same level; and the velocities at any two points, not on the same level, are connected by the law that the difference of their squares is equal to the difference of levels multiplied by $2\,g$.

40. Composition of Motions.—Principle 2 of last section is a particular case of the following general law:—

When a force acts upon a body already in motion, the subsequent motion will be obtained by compounding (in the same manner as forces are compounded by the parallelogram of forces) the motion which the force would have imparted to the body if initially at rest, with the motion which the body would have had in the absence of the force. The law, as thus stated, is applicable without qualification as long as the force continues to act parallel to a definite direction. In the case of forces which do not fulfil this condition, but gradually change their direction in space, the motion may be approximately determined by dividing the time into intervals so short, that the direction of the force does not change by a sensible amount during any one interval. The motion in each interval can then be determined by the above law. It is frequently possible, by the aid of the higher mathematics, to foresee the exact result to which this tedious method would only approximately lead, but the physical principles on which the investigation is conducted are in all cases those which we have above indicated.

41. Uniform Acceleration.—The general law above stated is exemplified in the case of gravity. In fact, if a denote the space traversed in the first second of a falling body's descent, the space in two seconds is $4a$, and consequently the space traversed in the second second is $3a$. Now the velocity at the end of the first second is $2a$, and this velocity, if it remained unchanged, would cause the space $2a$ to be traversed in the next second. Hence the space $3a$ actually described may be divided into two parts, $2a$ and a, of which the latter is due to the continued action of the force of gravity during the second second.

In like manner the space described in the third second is $9a - 4a$, that is, $5a$, which may be divided into two parts, $4a$ and a, of which the latter alone is due to the action of gravity during the third second, the former being due to the velocity $4a$, which the body possessed at the end of the preceding second.

So, again, the velocity acquired by the body in the first second being $2a$, the velocity at the end of any subsequent second will be found to be the sum of two parts, one of which is the velocity at the beginning of the second, and the other is $2a$.

Motion possessing these properties is said to be *uniformly accelerated;* and the force which produces it is a uniformly accelerating force, that is to say, a constant force.

The force of gravity, however, is sensibly constant only within moderate limits of distance from the earth's surface; as we ascend from the earth its intensity continually diminishes, being nearly proportional inversely to the square of the distance from the earth's centre. Hence a body falling *in vacuo* from a great height towards the earth, would not be uniformly accelerated, but would experience continually greater acceleration as it descended. On the other hand, if there were a vacuous shaft down which a body could fall to the centre of the earth, it would fall with continually diminishing acceleration, because the force of gravity in the interior of a solid sphere diminishes as we approach the centre, and becomes zero at the centre itself, where the attractions, being equal in all directions, destroy one another. A body so falling would have its velocity continually increased, but the *rate of increase as measured by the difference between the velocity at the beginning of a second and that at the end of it,* would continually become less. The words italicized in last sentence constitute the definition of *acceleration.* It denotes the rate of increase of velocity, just as velocity itself denotes the rate of increase of distance measured along the path described from a fixed point.

42. **Proportionality of Acceleration to Force directly and to Mass inversely.**—The general law of motion enunciated in § 40 may be extended to the case of several forces acting simultaneously. The actual motion will be obtained by compounding (on the parallelogram principle) the motions due to the separate forces together with the motion (if any) due to the initial velocity. Just as two forces acting on a point in the same direction are equivalent to a single force equal to their sum; so two motions in the same direction constitute, when compounded, a motion equal to their sum; and this is true, both as regards velocity and space described. Two equal forces acting on a body in the same direction, will therefore produce in any given time double the velocity that one of them would have produced alone. We are thus led to the general principle, that the velocities produced in equal bodies by different forces are simply proportional to the forces.

The above reasoning is not offered as an *à priori* proof of the general principle in question, but as a logical deduction of it from the law of composition of motions due to several forces. The proportionality of velocity to the force which produces it can be proved experimentally by Attwood's machine, and in other ways; and the law of composition in question must be regarded as established by the experimental verification of these and other consequences to which it leads.

From the direct proportionality of velocity to the force by which it is generated, when the mass is given, we may infer the inverse proportionality of velocity to the mass which is set in motion, when the force is given. For instance, if we double the mass, leaving the force unchanged, we may resolve the force into two equal parts, acting one on each half of the mass. Doubling the mass has, therefore, the same effect on the motion as halving the force.

The velocity generated in a given time is thus proportional to the moving force divided by the mass moved; from which it follows that force is proportional to the product of mass by velocity generated in a given time.

When force is expressed in terms of the "absolute" or "invariable" unit, first proposed by Gauss, we can assert that the moving force is *equal* to the product of the mass moved, and the velocity generated in a unit of time.[1] For example, since the force of gravity on a body weighing M pounds causes it to acquire a velocity of g feet in a second, this force is numerically equal to Mg, it being understood that the pound is the unit of mass, and the foot and second the units of space and time.

That the pound is really and strictly a standard of mass is obvious from the consideration that the standard pound is a certain piece of platinum, preserved at the office of the Exchequer in London, and that this piece of platinum would remain a true pound if carried to any part of the earth.

At any one place, since g has a given value, the masses of bodies are proportional simply to their weights, but this proportion obviously does not hold in the comparison of masses at places where the values of g are different. When a body is carried about to different parts of space, its mass, or quantity of matter, remains of course the same,

---

[1] The absolute unit of force may be defined as that force which, acting on unit mass for unit time, would generate unit velocity. The force of gravity on unit mass contains g such units.

but its weight alters in proportion to the greater or less intensity of gravity;—for instance, at the centre of the earth, regarded as a uniform sphere, its weight would be nothing, This annihilation of its weight would in no way affect its resistance to acceleration. The difference between the mass of a ball of cork, and that of a ball of lead of the same diameter, could in such circumstances be readily detected by the different resistances which they would oppose to attempts to set them in rapid motion, or to check their motion when commenced.

It must be regarded as a remarkable fact, and one which could only have been established by experiment, that the two modes of comparing masses perfectly coincide; that is to say, two bodies, even though composed of different materials, if their sizes are so proportioned that they oppose equal resistances to acceleration, will also gravitate with equal forces, as tested by their balancing each other in a pair of scales. This principle is established experimentally by the equal velocities of fall of all bodies in vacuo, and, with much greater accuracy, by the equality of the number of vibrations made in the same time by pendulums of the same size and form, but of different materials.

CHAPTER VI.

43. The Pendulum.—When a body is suspended so that it can turn about a horizontal axis which does not pass through its centre of gravity, its only position of stable equilibrium is that in which its centre of gravity is in the same vertical plane with the axis and below it (§ 28). If the body be turned into any other position, and left to itself, it will oscillate from one side to the other of the position of equilibrium, until the resistance of the air and the friction of the axis gradually bring it to rest. A body thus suspended, whatever be its form, is called a pendulum. It frequently consists of a rod which can turn about an axis O at its upper end, and which carries at its lower end a heavy lens-shaped piece of metal M called the bob; this latter can be raised or lowered by means of the screw V. The applications of the pendulum are very important: it regulates our clocks, and it has enabled us to measure the intensity of gravity and ascertain the differences in its amount at different parts of the world; it is important then to know at least the fundamental points in its theory. For explaining these we shall begin with the consideration of an ideal body called the *simple pendulum.*

44. Simple Pendulum.—This is the name given to a pendulum consisting of a heavy particle M attached to one end of an inextensible thread without weight, the other end of the thread being fixed at A. When the

Fig. 34.—Pendulum. thread is vertical the weight of the particle acts in the direction of its length, and there is equilibrium. But suppose it is drawn aside into another position, as AM. In this case the weight MG of the particle can be resolved into two forces MC and MH. The former, acting along the prolongation of the thread, is destroyed by the resistance of the thread; the other, acting along the tangent MH, produces the motion of the particle. This effective component is evidently so much the greater as the angle of displacement from the vertical position is greater. The particle will therefore move along an arc of a circle described from A as centre, and the force which urges it forward will continually diminish till it arrives at the lowest point M'. At M' this force is zero, but, in virtue of the velocity acquired, the particle will ascend on the opposite side, the effective component of gravity being now opposed to the direction of its motion;

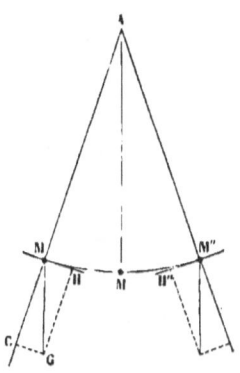

Fig. 35.—Motion of Simple Pendulum.

and, inasmuch as the magnitude of this component goes through the same series of values in this part of the motion as in the former part, but in reversed order, the velocity will, in like manner, retrace its former values, and will become zero when the particle has risen to a point M" at the same height as M. It then descends again and performs an oscillation from M" to M precisely similar to the first, but in the reverse direction. It will thus continue to vibrate between the two points M, M" (friction being supposed excluded), for an indefinite number of times, all the vibrations being of equal extent and performed in equal periods.

The distance through which a simple pendulum travels in moving from its lowest position to its furthest position on either side, is called its *amplitude*. It is evidently equal to half the complete arc of vibration, and is commonly expressed, not in linear measure, but in degrees of arc. Its numerical value is of course equal to that of the angle MAM', which it subtends at the centre of the circle.

The *complete period* of the pendulum's motion is the time which it occupies in moving from M to M" and back to M, or more generally, is the time from its passing through any given position to its next passing through the same position *in the same direction*.

What is commonly called the time of vibration, or the time of a single vibration, is the half of a complete period, being the time of passing from one of the two extreme positions to the other. Hence what we have above defined as a complete period is often called a double vibration.

When the amplitude changes, the time of vibration changes also, being greater as the amplitude is greater; but the connection between the two elements is very far from being one of simple proportion. The change of time (as measured by a ratio) is much less than the change of amplitude, especially when the amplitude is small; and when the amplitude is less than about 5°, any further diminution of it has little or no sensible effect in diminishing the time. *For small vibrations*, then, *the time of vibration is independent of the amplitude.* This is called the law of *isochronism*.

The time of a single vibration when the amplitude is small is expressed by the formula:

$$T = \pi \sqrt{\frac{l}{g}},$$

l denoting the length of the pendulum, g the intensity of gravity, and π the ratio of the circumference of a circle to the diameter. As regards the units in which T, l, and g are expressed, it must be remarked that if g is expressed in the usual way, as in § 38, l must be expressed in feet, and the value obtained for T will be in seconds.

The formula shows that the time of vibration is proportional to the square root of the length of the pendulum, so that if the pendulum be lengthened four, nine, or sixteen fold, the time will be doubled, trebled, or quadrupled.

45. Experimental Laws of the Motion of the Pendulum.—The preceding laws apply to the simple pendulum; that is to say, to a purely imaginary existence; but they are approximately true for ordinary pendulums, which in contradistinction to the simple pendulum are called compound pendulums. The discovery of the experimental laws of the motion of pendulums was in fact long anterior to the theoretical investigation. It was the earliest and one of the most important discoveries of Galileo, and dates from the year 1582, when he was about twenty years of age. It is related that on one occasion, when in the cathedral of Pisa, he was struck with the regularity of the oscillations of a lamp suspended from the roof, and it appeared to him that these oscillations, though diminishing in extent, preserved the same duration. He tested the fact by repeated trials, which confirmed him in the belief of its perfect exactness. This law of isochronism can be easily verified. It is only necessary to count the vibrations which take place in a given time with different amplitudes. The numbers will be found to be exactly the same. This will be found to hold good even when some of the vibrations compared are so small that they can only be observed with a telescope.

The time of vibration, then, does not depend on the amplitude, and neither does it depend on the material of which the pendulum is composed. From this last fact it follows that gravity acts in precisely the same manner on all substances. It is found, in fact, that balls of the same size, of lead, copper, ivory, &c., suspended by threads of equal length, vibrate in the same time, provided they are large enough to escape sensible retardation from the resistance of the air. This result is virtually identical with that of Galileo's experiment on the fall of bodies (§ 32), and enables us to conclude with certainty that in a vacuum these different pendulums would vibrate in rigorously equal times.

By employing balls suspended by threads of different lengths, Galileo discovered the influence of length on the time of vibration. He ascertained that when the length of the thread increases, the time of vibration increases also; not, however, in proportion to the length simply, but to its square root.

Knowing, then, that the length of the pendulum which beats seconds at Paris is about one metre ($0^m.994$), we see that a pendulum 64 metres long would make a single vibration in eight seconds. This is about the length of the pendulum employed by Foucault at the Pantheon in his celebrated experiments on the rotation of the earth.

This law of lengths experimentally discovered by Galileo, is precisely that which the formula for the simple pendulum gives; an agreement which it was natural to expect, seeing that a small ball suspended by a long string is a practical approximation to the idea of a simple pendulum. When, however, the form of the pendulum departs widely from this, the meaning to be attached to the word *length* ceases to be obvious. It becomes necessary to resort to theoretical investigations, and we shall briefly indicate the results thus obtained.

46. **Equivalent Simple Pendulum.**—It is demonstrated in treatises on dynamics that a compound pendulum, whatever be its form,

always keeps time with a simple pendulum of some determinate length—called the *equivalent simple pendulum;*[1] and whenever the length of a pendulum is mentioned, it is the length of the equivalent simple pendulum that is to be understood.

Any rigid body, oscillating about a fixed horizontal axis, may be regarded as a compound pendulum. That point of the axis which, in the position of equilibrium, is vertically over the centre of gravity of the body, is called the *centre of suspension;* and if we join this point to the centre of gravity, and produce the joining line downwards till its whole length is equal to that of the equivalent simple pendulum, its lower extremity is called the *centre of oscillation.* The body therefore vibrates in the same manner as if its whole mass were collected at the centre of oscillation. This point is always further from the axis than the centre of gravity, and it possesses the following remarkable property:—that if the body were made to vibrate about an axis passing through the centre of oscillation and parallel to the original axis, the time of vibration would be the same as in vibrating about the original axis. In this inverted position, the original centre of suspension becomes the new centre of oscillation, and the original centre of oscillation becomes the new centre of suspension; hence the property in question is commonly called the *convertibility of the centres of oscillation and suspension.*[2]

This important property, which was discovered by Huyghens, furnishes an accurate method of determining the length of the simple pendulum equivalent to a given compound pendulum. This is the principle of Kater's pendulum, which can be made to vibrate about either of two parallel knife-edges, one of which can be adjusted to any distance from the other. The pendulum is swung first upon one of these edges and then upon the other, and, if any difference is detected in the times of vibration, it is corrected by moving the adjustable edge. When the difference has been completely destroyed,

---

[1] Sometimes the *isochronous* simple pendulum; but it seems better to reserve this adjective and the corresponding noun isochronism for the use in which the latter has been employed in § 44.

[2] The names *centre of suspension* and *centre of oscillation* are not very appropriate, inasmuch as the properties mentioned in the text are properties of two lines rather than of two points: one of the lines being the axis about which the pendulum swings, and the other being a parallel to this axis in the plane containing it and the centre of gravity, and at a distance from it equal to the length of the equivalent simple pendulum. If we call the former the *axis of suspension* and the latter the *axis of oscillation,* we may assert that the axes of suspension and oscillation are convertible.

tne distance between the two edges is the length of the equivalent simple pendulum. It is necessary, in any arrangement of this kind, that the two knife-edges should be in a plane passing through the centre of gravity; also that they should be on opposite sides of the centre of gravity and at unequal distances from it.

We see, by what precedes, that the laws of the simple pendulum are applicable to any pendulum, if we understand by its length the length of the equivalent simple pendulum, or, in other words, the distance between the axis and the centre of oscillation.

47. Determination of the Value of g.—Returning to the formula for the simple pendulum $T = \pi \sqrt{\dfrac{l}{g}}$, we easily deduce from it $g = \dfrac{\pi^2 l}{T^2}$, whence it follows that the value of g can be determined by making a pendulum vibrate and measuring T and l. T is determined by counting the number of vibrations that take place in a given time; l can be calculated, when the pendulum is of regular form, by the aid of formulæ which are given in treatises on rigid dynamics, but its value is more easily obtained by Kater's method, described above, founded on the principle of the convertibility of the centres of suspension and oscillation.

48. Variations in the Intensity of Gravity.—Pendulum observations, which have been taken in great numbers in various places, have established the result that the intensity of gravity varies over the surface of the earth. At London the value of g is 32·182; it increases in approaching the pole, and diminishes in going towards the equator. These variations, however, are not very considerable, as the following table of the values of g shows:—

Value at the equator....................32·088.
Value in latitude 45°....................32·171.
Value at the poles........................32·253.

Its value for any place is approximately given by the formula:—

$$g = 32\cdot088\ (1 + \cdot005133 \sin^2\!\lambda) \left(1 - \frac{2h}{R}\right),$$

λ denoting the latitude of the place, h the height above the level of the sea,[1] and R the earth's radius, which is 20,900,000 feet. Local

---

[1] The correcting factor for elevation $\left(1 - \dfrac{2h}{R}\right)$ is proper to be used in determining the value of g in mid-air; for example, in the positions reached by balloons. On the summit of a mountain, or of an elevated plateau in the interior of a continent, the value of g is greater than at the same level in mid-air above the ocean, owing to the attraction of the excess of land which projects above sea-level. See a paper by Dr. Young in the *Phil. Trans.* for 1819.

peculiarities prevent the possibility of laying down any general formula with precision, and the exact value of g for any place can only be ascertained by observations on the spot.

49. Centrifugal Force.—There are two distinct causes of these differences in the intensity of gravity: the first is what is called *centrifugal force.* Consider a material point M attached to the extremity of a thread OM, and suppose an impulse to be given it in a direction perpendicular to the thread. At each instant, in virtue of inertia, the point tends to move along a tangent to the circle; but the thread prevents this movement from taking place, by drawing the point, which in its turn reacts on the thread and stretches it with a certain force which has received the name of centrifugal force. It is clear

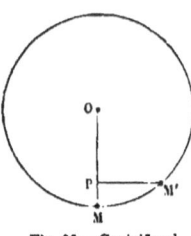

Fig. 36.—Centrifugal Force.

that we may dispense with the thread, if we suppose an attractive force equal to that which the thread would exert directed towards O. This force, which, by combining its effect with that of the initial impulse, would produce circular motion, is called *centripetal force.* It is evidently equal and opposite to centrifugal force.

The amount of the centripetal force in any given case of circular motion can be easily calculated. Let MP represent the space which the material point would describe in a certain small time t under the action of the centripetal force alone. This motion, combined with the motion due to the velocity which the particle possessed at M, gives the actual motion in the arc MM', provided that the time considered and the distance moved in that time are so small that the directions of the centripetal forces at M and M' are sensibly parallel—in other words, provided the arc MM' is very small in comparison with the radius OM. In this case, by § 40, PM', or rather a line equal and parallel to it drawn through M, represents the space due to the velocity which the point had at M, and is therefore equal to vt, v denoting the velocity. Now if we denote by ϕ the acceleration due to the centripetal force, this acceleration may be regarded as uniform during the time t, and we have by § 38, $MP = \frac{1}{2}\phi t^2$, whence $\phi = \frac{2MP}{t^2}$. But PM'^2, that is v^2t^2, is (by *Euclid,* iii. 35) equal to $2r.MP$, r denoting the radius of the circle; therefore $\frac{2MP}{t^2}$ is equal to $\frac{v^2}{r}$. Consequently $\phi = \frac{v^2}{r}$; that is, the centripetal force upon a particle revolving with velocity v (feet per second) in a circle of radius r (feet) is equal to a force which, acting continuously upon the same particle initially at rest, would in one second give it a velocity of $\frac{v^2}{r}$. The centripetal force upon a mass of m pounds is $\frac{mv^2}{r}$ Gaussian pound units (42), and is equal to the weight of $\frac{mv^2}{gr}$ pounds.

When the circular movement is uniform, we can give the formula for φ a more convenient form. In fact if we denote by T the time of revolution, and by π the ratio of circumference to diameter, we have $v = \frac{2\pi r}{T}$, and $\varphi = \frac{v^2}{r} = \frac{4\pi^2 r}{T^2}$.

Suppose, for example, a weight of 50 pounds attached to one end of a string 3 feet long, and revolving uniformly in a circle about the other end of the string, at the rate of 40 revolutions per minute. The time of revolution is here $\frac{3}{2}$ of a second, and the force required to be exerted by the string is equal to the weight of

$$\tfrac{50}{32\cdot 2} \times 4\pi^2 \times \tfrac{4}{9} \times 3 = 81\cdot 7 \text{ pounds};$$

and if the string be not strong enough to bear this weight it will break, and the body will fly off at a tangent.

50. Different Effects of Centrifugal Force.—Several experiments on centrifugal force are exhibited in courses of physics. For example, a rod AB (Fig. 37), passing through two ivory balls M, M', is set in rotation in a horizontal plane by means of the mechanism shown in the figure. The balls are then seen to move towards the extremities of the rod. If a spring is placed beside one of the balls M', as shown in the figure, it will be pressed with a force which is precisely equal to the centrifugal force of the ball.

The centrifugal railway (Fig. 38) shows a curious effect of this force. A carriage, starting from A, descends the inclined rails, and by following the course of the rails, which here forms a spiral convolution, to B, rises to C, and descending again on the side next A, passes on the further side of B and finally arrives at D. There is therefore an instant in the motion when the carriage is bottom upwards at the top of the convolution, and remains in contact with the rails in opposition to the force of gravity. The explanation is easy. The carriage attains a certain velocity in descending the incline which forms the first portion of its path. In virtue of its inertia it tends to move with this velocity in a tangential direc-

tion, but being compelled to follow the curved rails which lie in
its way, it reacts upon them with a force whose amount is given

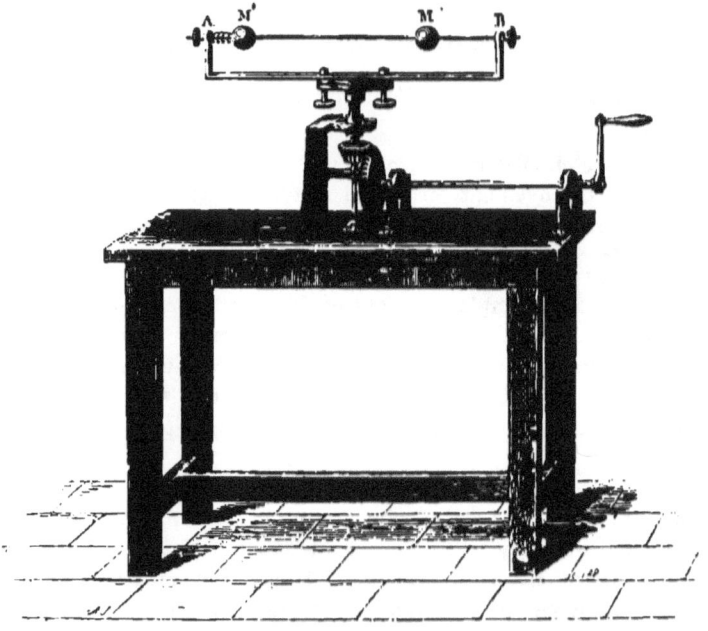

Fig. 37.—Centrifugal-force Apparatus.

by the formulæ of last section. If the point A is sufficiently elevated
above C, this force will prevail over the weight of the carriage

Fig. 38.—Centrifugal Railway.

and keep it pressed against the rails. It may be easily shown that
to secure this result the height of A above B must be to the height
of C above B in a greater ratio than 5 to 4.

If the apparatus shown in Fig. 39, consisting of a kind of sphere formed of four flexible springs, be mounted on the whirling table and made to revolve, it will be seen to become flattened, the effect being more decided as the rotation is more rapid. This result is due to centrifugal force, which gives all parts of the springs a tendency to recede from the axis of rotation, and especially those parts which are already most distant from it. This experiment may be regarded as illustrating the manner in which the earth, when in a fluid' state, acquired its present form, bulging at the equator and flattened at the poles.

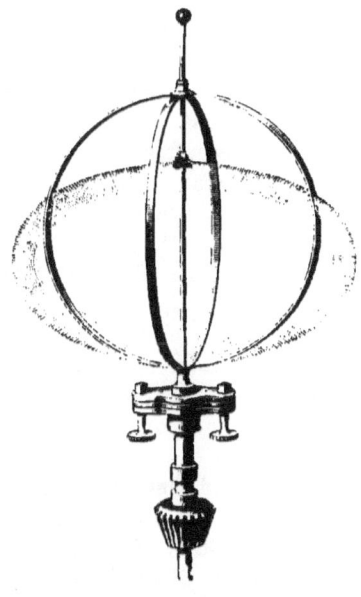

89. Oblateness of the Earth.

51. The influence of centrifugal force in modifying the effect of gravity is easily deduced from what precedes. The various bodies which are on the surface of the earth are retained upon it by gravity, and a certain portion of the force of gravity is expended in constraining them to move in circular paths. The modification thus produced in the apparent force and direction of gravity is the same as if a force equal and opposite to the force thus expended were compounded with the force of gravity proper; that is to say, apparent gravity is the resultant of gravity proper and centrifugal force.

At the equator, centrifugal force is directly opposed to gravity proper, and is therefore to be simply subtracted from it. Let r be the earth's radius, T the length of a sidereal day (or the time of the earth's revolution), then the intensity of centrifugal force at the equator is $\frac{4\pi^2 r}{T^2} = \frac{2\pi r \cdot 2\pi}{T^2}$. Putting for $2\pi r$, or the earth's circumference, its value in metres, viz. 40,000,000, and for T its value in seconds, 86,164, we obtain for the intensity in question the value ·033. Now the intensity of gravity, expressed with reference to the metre and second as units, is about 9·8. Hence centrifugal force at

5

the equator is $\frac{1}{289}$ or about $\frac{1}{200}$ of the force of gravity, and apparent gravity is less than gravity proper at the equator by this amount. The exact amount of diminution is more nearly $\frac{1}{289}$ of gravity proper; and since 289 is the square of 17, and centrifugal force varies as the square of the velocity, we see that if the rate of the earth's revolution were increased seventeenfold, bodies at the equator would lose their weight.

As we recede from the equator, the intensity of centrifugal force diminishes because the distance from the earth's axis diminishes; and, at the same time, the direction of centrifugal force, being always perpendicular to this axis, becomes less directly opposed to gravity, so that only one of its two components is subtractive. For this double reason, the effect of centrifugal force in diminishing the apparent force of gravity becomes continually less as we recede from the equator. As far then as the disturbing effect of centrifugal force is concerned, the apparent intensity of gravity should be least at the equator, and should continually increase towards the poles.

We may add that centrifugal force (except at the equator) affects not only the amount, but also the direction, of apparent gravity, since the resultant of two forces which are not directly opposed does not coincide in direction with either of them. The angle which the actual vertical (indicating the direction of apparent gravity), makes with the direction which the vertical would have if the earth were at rest, varies with the latitude; at Paris, where its value is nearly a maximum, it is between 5 and 6 minutes.

52. Universal Gravitation, Earth's Mean Density.—Terrestrial gravitation is only a particular case of universal gravitation. Newton established, by researches extending over more than twenty-five years, that the movements of the planets around the sun, and of the satellites around the planets, could be explained by assuming the existence of a mutual attraction, which, taken in conjunction with an initial impulse, determines the paths which these bodies describe. This attraction is jointly proportional to the masses of the mutually attracting bodies, and varies inversely as the square of their mutual distance. By the aid of this assumption, astronomers have succeeded not only in explaining all the diversities of motion which the solar system exhibits, but in calculating the positions of the celestial bodies at distant times, both past and future, with marvellous accuracy. The study of the movements of the heavenly bodies, which has thus been reduced to a branch of applied mathematics, is called Physical Astronomy.

It is therefore natural to look upon terrestrial gravity as a particular case of universal gravitation, and to regard the fall of a body as a consequence of the attraction exerted on it by the different parts of the terrestrial globe. This identity is in fact established by numerous proofs. Now it is obvious from the symmetry of the component attractions, that the resultant attraction of a sphere exerted at any point of its surface must be in the direction of the radius, and must therefore be perpendicular to the surface—a result which our every-day experience confirms in the case of the earth's gravitation. Theory also shows that the attraction of an oblate spheroid is greater at the poles than at the equator. In fact, taking into account both the oblateness of the earth and centrifugal force, theory agrees with observation in indicating that apparent gravity increases in going from the equator towards the poles by an amount proportional to the square of the sine of the latitude.

The hypothesis of attraction then explains all the phenomena of gravity: we may add that the existence of attraction at the surface of the earth has been experimentally demonstrated. Maskelyne, Hutton, and Playfair, in the celebrated Schiehallien experiment, proved that the plumb-line experienced, on either side of the mountain of that name, a deviation towards the mountain; and a similar result was established by Sir Henry James at Arthur's Seat, near Edinburgh. Cavendish, by means of a very sensitive torsion-balance, showed that two large spheres of lead exercised a sensible attraction upon two small spheres, and was able to measure the amount of this attraction so precisely as to deduce, from the comparison of it with the earth's attraction, that the mean specific gravity of the earth is 5·5. Cavendish's experiment has been repeated by Reich in Germany and by Baily in this country, with nearly coincident results. The Schiehallien experiment, which was the earliest attempt to determine the earth's mean density, gave 5 for the result, and Mr. Airy's experiment at Harton Colliery gave a result exceeding 6. Baily's result, obtained by Cavendish's method, with some improvements in the details, is generally accepted as the best determination; it is 5·67.

53. Variation of Gravity with Height.—It follows from the identity of attraction and gravity that this force diminishes as we rise above the earth's surface, but the heights with which we commonly have to do in our experiments are so small in comparison with the earth's radius that the resulting differences of force are quite inappreciable. In the case, however, of large differences of height, the variation of force can be detected; for instance, it is easy to establish experimentally that the intensity of gravity is less on the summit of a mountain than at its base. Theory shows that the attraction of a sphere upon external points is the same as if its mass were collected at the centre. Presuming this to be true in the case of the earth, it may be shown that, in ascending to any height above the earth's surface which is small in comparison with the earth's radius, the diminution in the force of gravity is twice as great in comparison with the whole force of gravity as that height is in comparison with the earth's radius. This explains the origin of the factor $1 - \frac{2h}{R}$ in the formula of § 48.

In penetrating into the interior of the earth, the law of the variation of gravity is more complex. If the earth were homogeneous, its attraction would continually diminish in penetrating towards the centre, and would, if the earth were also truly spherical, be simply proportional to the distance from the centre. But in fact the density is greater in the central than in the superficial parts of the earth, the mean density being, as we have seen, about 5·7, while the density of the superficial beds is only about 3. This augmentation of density tends to increase the attractive force as we descend; and it will depend upon the law according to which the density varies which of these two opposite tendencies will prevail. Mr. Airy, in his experiment at Harton Colliery, found the intensity of gravity at the bottom of the mine, 1256 feet below the surface, to be greater than at the surface by about one part in 19,000. Supposing that similar results would be obtained in other places, it follows that, in penetrating the earth, gravity must go on increasing from the surface down to a certain depth, where it becomes a maximum, and from which it decreases down to the earth's centre, where it becomes zero, the equal and opposite attractions there destroying one another.

53 A. Simple Vibrations.—The motion of a pendulum vibrating in a small arc, may be taken as the type of a class of motions which very extensively prevail in nature, and are of great importance in many departments of physical science. They have been called by different writers *simple vibrations, pendulum-like vibrations, sine-like vibrations,* and *simple harmonic motions.* We shall employ the first of these designations, and define a simple vibration to be the movement of a point to and fro along one line (not necessarily straight), with acceleration simply proportional to the distance of the point from the bisection of this line. If the line of motion be curved, the distance is to be measured along the curve, and in this case the acceleration referred to is the tangential acceleration along this curve. In every case, and in all parts of the motion, the acceleration of the moving point urges it *towards* the middle point of its path, becoming zero only at the instant of passing this middle point. In the majority of cases that require to be considered, the line of motion is straight, or approximately straight.

As examples of simple vibrations, we may instance, in addition to that of a pendulum above quoted, the motion of a point on either prong of a tuning-fork, or of a point in a musical string, when vibrating so as to produce a pure note. In these cases, the force urging any point of the fork or string towards the position of equilibrium, varies directly as the distance of the point at any time from this position, and by the second law of motion acceleration is proportional to force.

In the case of the simple pendulum (§ 43), the acceleration of the heavy particle when the thread makes an angle θ with the vertical is $g \sin \theta$ (see § 53D), which when the arc of vibration is small may be taken as equal to $g\theta$, that is, $g\,\dfrac{x}{l}$, x denoting the length of arc

measured from lowest point, and l the length of the string. It is therefore proportional to x, the distance of the particle from the position of equilibrium.

53 B. When simple vibration is executed in a straight line, it corresponds to the projection of uniform circular motion (Fig. 39A); that is to say, if a point P move with uniform velocity round the circumference of a circle, and if Pp be the perpendicular let fall from P upon a fixed straight line in the plane of the circle, then p, the foot of this perpendicular, will execute simple vibration.

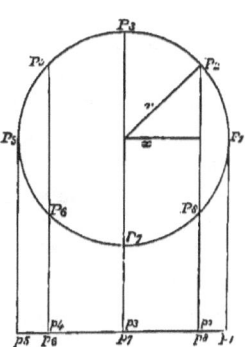

Fig. 39A. Projection of Circular Motion.

To prove this proposition, remark that by § 49 the point P is to be regarded as constantly falling away from a tangent in obedience to acceleration directed towards the centre of the circle, the amount of the acceleration being $\frac{4\pi^2}{T^2}.r$, where r denotes the radius of the circle. This acceleration, being directed from P towards the centre of the circle, may be resolved into two components, one parallel to the line of motion of p, and the other perpendicular to it. The former of these, which is obviously the acceleration of p, is $\frac{x}{r}$ of the whole acceleration, x denoting the distance of p from its mean position, that is, from the foot of a perpendicular let fall from the centre of the circle. The acceleration of p is therefore $\frac{4\pi^2}{T^2}x$, and is proportional to x.

53 c. We may hence prove that simple vibrations are isochronous. For if the acceleration be expressed by μx, x denoting displacement from mean position, and μ any constant, we see, by the preceding paragraph, that the vibration keeps time with an imaginary point moving uniformly round a circle, a complete vibration being performed in the same period as a complete revolution; and if T denote this period, we have $\mu = \frac{4\pi^2}{T^2}$; whence $T = \frac{2\pi}{\sqrt{\mu}}$, an expression which is independent of the amplitude of vibration.

Applying this result to a pendulum vibrating in a small arc, we have $\mu = \frac{g}{l}$ (see § 53A); hence $T = \frac{2\pi}{\sqrt{\mu}} = 2\pi\sqrt{\frac{l}{g}}$, which is the time of a complete (or double) vibration.[1]

To understand the reason of the isochronism of simple vibrations, we have only to remark that, if the amplitude be changed, the velocity at corresponding points (that is, points whose distances from the middle point are the same fractions of the amplitudes) will be changed in the same ratio. For example, compare two simple vibrations in which the values of μ are the same, but let the amplitude of

---

[1] The mathematical reader will remark that our definition of simple vibration corresponds to the differential equation $\frac{d^2x}{dt^2} = -\mu x$, and that the property proved in § 53B corresponds to the solution of this equation, which is $x = a \cos(t\sqrt{\mu} - \epsilon)$, a and ϵ being arbitrary constants, the former called the *amplitude* and the latter the *epoch* of the motion. The quantity $(t\sqrt{\mu} - \epsilon)$ is called the argument, and it is obvious that x and its successive differential coefficients will all remain unaltered if the argument be increased by 2π, that is, if t be increased by any multiple of $\frac{2\pi}{\sqrt{\mu}}$. Hence the motion always repeats itself after the interval $\frac{2\pi}{\sqrt{\mu}}$, which is therefore the *period* of a complete vibration.

one be double that of the other. Then if we divide the paths of both into the same number of small equal parts, these parts will be twice as great for the one as for the other; but if we suppose the two points to start simultaneously from their extreme positions, the one will constantly be moving twice as fast as the other. The number of parts described in any given time will therefore be the same for both.

In the case of vibrations which are not simple, it is easy to see (from comparison with simple vibration) that if the acceleration increases in a greater ratio than the distance from the mean position, the period of vibration will be shortened by increasing the amplitude; but if the acceleration increases in a less ratio than the distance, as in the case of the common pendulum vibrating in an arc of moderate extent, the period is increased by increasing the amplitude.

53 D. Cycloidal Pendulum.—We saw in § 33 that the effective component of gravity upon a particle resting on a smooth inclined plane was proportional to the sine of the inclination. The acceleration of a particle so situated is in fact $g \sin a$, if a denote the inclination of the plane. When a particle is guided along a smooth curve its acceleration is expressed by the same formula, a now denoting the inclination of the curve at any point to the horizon. This inclination varies from point to point of the curve, so that the acceleration $g \sin a$ is no longer a constant quantity. The motion of a common pendulum corresponds to the motion of a particle which is guided to move in a circular arc; and if x denote distance from the lowest point, measured along the arc, and r the radius of the circle (or the length of the pendulum), the acceleration at any point is $g \sin \dfrac{x}{r}$. This is sensibly proportional to x so long as x is a small fraction of r; but in general it is not proportional to x, and hence the vibrations are not in general isochronous.

To obtain strictly isochronous vibrations we must substitute for the circular arc a curve which possesses the property of having an inclination whose sine is simply proportional to distance measured along the curve from the lowest point. The curve which possesses this property is the cycloid. It is the curve which is traced by a point in the circumference of a circle which rolls along a straight line. The cycloidal pendulum is constructed by suspending an ivory ball or some other small heavy body by a thread between two cheeks (Fig. 39B), on which the thread winds as the ball swings to either side. The cheeks must themselves be the two halves of a cycloid whose length is double that of the thread, so that each cheek has the same length as the thread. It can be demonstrated[1] that under these circumstances the path of the ball will be a cycloid identical with that to which the cheeks belong. Neglecting friction and the rigidity of the thread, the acceleration in this case is proportional to distance measured along the cycloid from its lowest point, and hence, by last section, the time of vibration will be strictly the same for large as for small amplitudes.

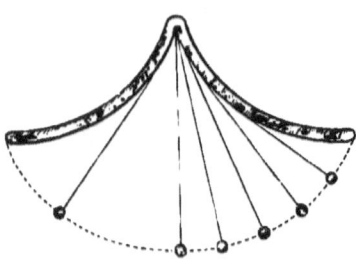

Fig. 39B.—Cycloidal Pendulum.

It will, in fact, be the same as that of a simple pendulum having the same length as the cycloidal pendulum and vibrating in a small arc.

Attempts have been made to adapt the cycloidal pendulum to clocks, but it has been found that, owing to the greater amount of friction, its rate was less regular than that of the common pendulum. It may be remarked, that the spring by which pendulums are often suspended has the effect of guiding the pendulum bob in a curve which is approximately cycloidal, and thus of diminishing the irregularity of rate resulting from differences of amplitude.

53 E. Centre of Mass, or Centre of Inertia.—The point which has been mentioned in Chapter iv., under the name of centre of gravity, possesses important properties besides those which depend upon its being the point through which the resultant force of gravity passes.

It may be demonstrated by geometry that if a body be divided into a number of equal elements (equal in mass, not necessarily in size), each of them being so small in all its dimensions that it may be treated as a material point, there is a certain point in the body such that its distance from any plane whatever is equal to the mean distance of all the elements from the same plane—that is, to the sum of all the distances divided by the number of elements.[2] This point is called the *centre of mass.*

---

[1] Since the evolute of the cycloid is an equal cycloid.

[2] If the plane cuts the body, distances on one side of the plane must be reckoned positive and on the other negative, and the sum in question must be the algebraical sum.

If a set of equal and parallel forces act one on each element in the same direction, their resultant will pass through the centre of mass. Inasmuch then as the force of gravity upon a body is made up of such a set of equal and parallel forces, the centre of mass is also the *centre of gravity.*

Again, if a body at rest be set in motion in such a manner that all its points move in parallel straight lines with equal velocities, the resistance which inertia opposes is composed of a set of such equal and parallel resistances; its resultant therefore passes through the centre of mass, which is hence called the *centre of inertia.* It is evident that the force requisite for producing such motion in a body must be equivalent to a single force applied at this point, and the same remark applies to the force necessary for destroying such motion and bringing the body to rest.

Conversely, if a force be applied to a body at its centre of mass, or in a line passing through the centre of mass, the body will be set in motion in such a way that all its points will have equal and parallel velocities, their common direction being parallel to the line of action of the force.

A force applied to a free body in a direction not passing through the centre of mass will produce movement of the centre of mass combined with rotation of the body about the centre of mass. Of these two components of motion, the former will be the same as would be produced by the given force if it acted in a direction passing through the centre of mass; and the latter—the rotation— will be the same as if the centre of mass were fixed.

A couple applied to a free body will produce rotation of the body about the centre of mass, but will not produce any motion of the centre of mass.

When a body moves so that all its points are at every instant travelling in the same direction (that is in parallel directions and towards the same parts) and with equal velocities, it is said to have a movement of translation. All straight lines in a body so moving remain always parallel to their original positions, and conversely; hence this property may be taken as the definition of movement of translation. Every possible motion of a rigid body can be resolved into motion of translation accompanied by motion of rotation, and the resolution can always be so effected that the axis of rotation at any instant shall be parallel to the direction of the movement of translation. It is always possible, and is generally convenient, to regard the motion of a rigid body under the action of any forces as compounded of a motion of translation of the body as a whole, and a rotation of the body about an axis passing through its centre of mass.

53 F. When two or more bodies, or parts of the same body, are free to move, it is impossible for any action exerted between them to alter the motion of their common centre of mass. It is also impossible for such action to alter the total angular momentum about the centre of mass. For example, when an animal is either jumping or falling, no movement that it can make in mid-air without touching other bodies can either alter the motion of its centre of gravity, or cause part of its body to rotate in one direction without causing the remainder to rotate in the opposite direction.

The recoil of fire-arms depends on the same principle. Whatever force the gases which are produced by the explosion of the powder exert in propelling themselves and the ball forwards, they must always exert the same force for the same time in urging the gun backwards. If a shell explodes at an elevation in the air, then, neglecting the effect of the wind, the common centre of gravity of the fragments of the shell and the products of explosion will describe the same path and with the same velocity which the centre of gravity of the shell would have had if there had been no explosion.

This principle is of great importance in the movement of the heavenly bodies. For example, neglecting any general movement which the solar system as a whole may have in space, we are entitled to assert that in whatever direction the common centre of gravity of the planets may be moving at any time, the centre of gravity of the sun must be moving in a parallel and opposite direction; inasmuch as the centre of gravity of the whole system, consisting of sun and planets, remains always at rest.

53 G. Moment of Inertia.—When a body is capable of turning about a definite axis, its inertia opposes resistance to any force which may be applied to set it in rotation, and, if it has once been set in rotation, its inertia gives it a tendency to continue rotating with constant velocity, so that it can only be brought to rest by the action of opposing force.

The power of a force as regards its tendency to produce rotation about an axis is called the *moment of the force about the axis*, and is measured by the *product of the force and the arm at which it acts*. If the body is acted on by more forces than one, the sum of the moments of the several forces about the axis is the measure of the total tendency to produce rotation, and is called the total moment of all the forces. It is to be understood that if some of the forces tend to make the body turn in one direction and others in the opposite direction, the moments of the one set must be reckoned positive and of the other negative, and the sum in question must be the algebraical sum.

On the other hand, the resistance which the inertia of the rotating body opposes to the action of forces tending to accelerate or retard its rotation is called its *moment of inertia*. The rate at which the angular velocity changes is equal to the total moment of the forces divided by the moment of inertia of the body.

The moment of inertia of a body about an axis is the sum of all the terms which are obtained by multiplying each element by the square of its distance from the axis.

The *angular momentum* of a rotating body is a name given to the product of the moment of inertia and the angular velocity. Equal forces acting at equal arms for the same time upon different bodies produce equal angular momenta.

The *energy of rotation* of a rotating body is half the product of its moment of inertia and the square of its angular velocity. Equal amounts of work spent upon different bodies in producing rotation yield equal amounts of energy of rotation.

These ideas may be illustrated by a reference to the use of fly-wheels in machinery. A fly-wheel is a wheel which, by means of its inertia, acts as an equalizer of the motion of the machine to which it is attached, resisting, to an extent measured by its moment of inertia, all sudden changes of velocity. It is chiefly employed in cases where either the driving power or the resistance to be overcome is liable to rapid alternations of magnitude. When the power is in excess of the resistance the motion of the fly-wheel is accelerated, and the energy thus accumulated is given out again when the resistance is in excess of the power, the inertia of the fly-wheel then assisting to overcome the resistance, while at the same time the velocity of the wheel is diminished.

Fly-wheels are always made with heavy rims, the rest of the wheel being usually as light as is compatible with the requisite strength. This arrangement is adopted with the view of obtaining the greatest possible moment of inertia; for if all the matter of the wheel were collected at its rim, the moment of inertia would be equal to the mass multiplied by the square of the radius.

53 H. Centre of Percussion.—We have already seen that when a force acts upon a rigid body in a direction not passing through the centre of mass, it tends to produce a motion consisting partly of translation and partly of rotation of the body about the centre of mass. This principle remains true when the force is applied in the shape of a blow, and may easily be tested experimentally in a rough way by suspending a straight rod by a long string attached to one end and striking it with a hammer in different points. If the rod be struck in a horizontal direction near its top, its bottom will at the instant of the blow move in the opposite direction, and if it be struck near the bottom the top will fly back. In each case there is some intermediate line at right angles to the direction of the blow, which neither moves forwards nor backwards at the instant of the blow, while points on opposite sides of it move in opposite directions. With reference to this line, regarded as an instantaneous axis of rotation, the point at which the body was struck is called the centre of percussion. It admits of proof that the centre of percussion with respect to any axis is the same as the centre of oscillation.

When a body is suspended so that it can rotate about an axis, if we desire to strike it without jarring the axis, it is necessary that the blow should be administered at the centre of percussion, and this remark is equally true if the body in question be the striking instead of the struck body. For example, the proper point of a bat for striking a ball so as not to jar the hands is the centre of percussion of the bat with respect to an axis passing through the hands.

53 I. Momentum, Energy of Motion.—The product of the mass and velocity of a body is called the momentum of the body. If equal forces act upon unequal masses *for the same time*, the momenta generated are equal. This principle applies to the recoil of fire-arms, supposing the gun to be free to move.

On the other hand, if equal forces act upon unequal masses originally at rest, *through equal distances* (and therefore do equal amounts of work upon them), the momenta generated will be unequal; the greater mass will receive the greater momentum. Equal products will however be obtained in this case, if we multiply each mass by the square of its velocity. In the case of a falling body, we have seen that the velocity acquired in falling through a height s is $v = \sqrt{2gs}$, whence $\frac{1}{2}v^2 = gs$, and if the mass of the body (in lbs.) is m, we have $\frac{1}{2}mv^2 = gms$. Now, the force which produces the descent is the weight of m lbs., which is equivalent to gm Gaussian

units of force, and as the space through which the force works is
s, the work done is gms, which is the second member of the above
equation, and is equal to $\frac{1}{2}mv^2$. We see, then, that in this case the
work done, expressed in Gaussian units of work (of which a foot-
pound contains g), is equal to half the product of the mass (in
pounds) and the square of the velocity. This principle is perfectly
general, and may be extended to bodies already in motion as well as
to bodies initially at rest by substituting for "$\frac{1}{2}mv^2$," "the change
produced in the value of $\frac{1}{2}mv^2$." Conversely, since the height to
which a body will rise when thrown upwards with a given velocity
is the same as the height from which it must fall to acquire that
velocity, it follows from the foregoing equations that the value of
$\frac{1}{2}mv^2$ at the commencement of the ascent is equal to the work which
gravity would do upon the body during its descent from the height
to which it rises to the point from which its ascent commenced; and
if we denote the product of force and distance moved in the case
when the direction of the motion is opposite to that of the force,
by the name *negative work*, we may assert that the diminution which
occurs in the value of $\frac{1}{2}mv^2$ during the whole ascent or during any
part of it is equal to the negative work done upon the body by
gravity during that part of the motion.

It is in this sense that work and motion are said to be convertible,
and the product $\frac{1}{2}mv^2$, whose changes of value are always equal
to the work done upon the body, is called the *energy of motion*, or
the *kinetic energy* of the body. This equality subsists not only for
the case of gravity, but for all forces whatever: we may assert universally (neglecting for the present the effects of friction and molecular changes), that when a body of mass m moves at one time with
a velocity v_1, and at a subsequent time with velocity v_2, the whole
amount of work done upon the body during the interval (the algebraic sum being taken if any of the work is negative) is equal to
$\frac{1}{2}mv_2^2 - \frac{1}{2}mv_1^2$.

The product $\frac{1}{2}mv^2$ has sometimes been called the *accumulated
work* in a body, or the work stored up in the body, inasmuch as a
moving body is able in virtue of its motion to overcome resistance
through such a distance that the work done (or product of resistance
and distance through which it is overcome) will be equal to $\frac{1}{2}mv^2$.
We have seen one example of this in the case of a body thrown
upwards, which overcomes the resistance mg of gravity through a
height s such that $mgs = \frac{1}{2}mv^2$.

53 j. Energy of Position, or Potential Energy.—We have now to introduce a new idea, which is of comparatively recent origin, and plays an important part in modern dynamics. When a body of mass m is at the height s above the ground, which we will suppose level, we can cause it to acquire a certain velocity v such that $\frac{1}{2}mv^2 = mgs$ by simply allowing it to fall to the earth. The position of a body in this instance confers the power to obtain motion, and therefore kinetic energy; and as we have just seen, kinetic energy can be made to yield work. A body in an elevated position may therefore be regarded as a reservoir of work: the water in a mill-dam is, in fact, a case in point; and for this reason such a body is said to possess *energy of position*, or, as it is more commonly called, *potential energy*. In contradistinction from this latter name, the energy which a moving body possesses in virtue of its motion is sometimes called *actual energy*.

It should be remarked that energy of position is essentially relative, depending on the position of one body with reference to one or more others. In the case just considered the other body is the earth. In order to be philosophically correct in our language, we should speak not of the potential energy of a body, but rather of the potential energy of two or more bodies with reference to each other in a given relative position; or more briefly, of the potential energy of a certain relative position of the bodies.

It must also be remarked that while we can speak with precision of the *difference* between the potential energies of two specified positions, we cannot in strictness assign a definite value to the potential energy of one specified position unless we know the limits to the possible motion of the bodies in obedience to their mutual forces. For example, in the case just considered—that of a body at a certain height above level ground—the present position of the body is compared with that which it will occupy when it lies upon the ground. But a shaft might be sunk in the ground, and with reference to the bottom of this shaft a body lying on the surface of the ground would possess a certain amount of potential energy, which must be added to that above considered to obtain the potential energy of the body in its first position as compared with the position which it would occupy when lying at the bottom of the shaft.

Whenever motion takes place in obedience to natural forces, the increase or diminution of potential energy which takes place in passing from one position to another is always exactly compensated by an opposite change in the total amount of kinetic energy; from which it follows that the sum of potential and kinetic energies remains unchanged. Whenever kinetic energy is increased at the expense of potential energy, the forces concerned do an amount of positive work equal to the amount by which the former is increased or the latter diminished. On the other hand, whenever potential energy is increased at the expense of kinetic energy, the forces do negative work equal in absolute value to the energy thus transferred. Instances of the former kind of transfer are furnished by the motion of a falling body and the motion of a planet from aphelion to perihelion; instances of the latter kind are furnished by the motion of a body thrown upwards, and the motion of a planet from perihelion to aphelion.

53 k. Effect of Friction upon Transformation of Energy.—Thus far we have been supposing that frictional resistances are neglected. Friction, in fact, causes an apparent loss of energy, but this loss is accompanied by a generation of heat which is itself a form of energy, and a definite amount of heat is produced by each unit of work thus apparently wasted. Conversely, whenever heat is employed as a motive power (in the steam-engine, for example), a quantity of heat is destroyed equivalent, on the same scale, to the work produced.

Another kind of energy is developed when friction is employed as a means of generating electricity. In this case the potential energy of electrical attraction which is called into existence is the precise equivalent of the work spent in producing it.

Similar principles apply to all other cases in which energy is apparently destroyed. *Any particular form of energy may be destroyed, but only on condition of an equivalent amount of energy in some other shape coming into existence.* The whole amount of energy in the universe cannot undergo either increase or diminution. This great natural law is called the *principle of the conservation of energy.*

The exact nature of the various forms of molecular energy, such as heat, light, electricity, magnetism, and chemical affinity, is not at present known, but we run little risk of error in affirming that they all consist either of peculiar kinds of molecular motion or of peculiar arrangements of molecules as regards relative position. They must therefore fall under one or other of the two heads "energy of position" and "energy of motion."

CHAPTER VII.

54. The object of the balance is the measurement of the weights of bodies. It consists essentially of a rigid lever AB called the beam, movable about an axis O at the centre of its length. This axis rests

Fig. 40.—Balance.

upon two planes, and as it is a little above the centre of gravity, the beam takes a position of stable equilibrium. An index needle attached to the beam traverses a graduated arc, and indicates the position of equilibrium of the beam by pointing to zero.

This equilibrium will not be disturbed if we suspend from the extremities of the beam two scale-pans of the same substance, form, and dimensions. Neither will it be disturbed if in these scale-pans we place bodies of equal weight. And conversely, if two bodies placed in the two scale-pans equilibrate each other, their weights are equal. This, then, is the principle of the well-known use of the balance.

55. Correctness of the Balance.—It is necessary to the validity of the preceding reasoning that the scale-pans should be suspended at exactly the same distance from the axis, or, in other words, that the arms of the balance should be rigorously equal in length. This is known to be the case if the needle points to zero both when the scale-pans are empty and when they are loaded with two bodies of equal weight. If we have not two weights exactly equal, it is sufficient to place any body whatever in one of the scale-pans, and equilibrate it by placing so much matter in the other scale as will bring the index to zero; if we then interchange the contents of the two scale-pans, the needle should still point to zero. If it does not, the reason is that the arms are not of equal length. Easy as it is, however, to make the arms of approximately equal length, it is exceedingly difficult to make them rigorously equal; and accordingly, whenever great accuracy is required, the method of double weighing is employed, which enables us to obtain the exact weight, even when the arms of the balance are slightly unequal. This method consists in first counterpoising the body to be weighed with any substance—as, for example, shot or sand—and then replacing the body by weights sufficient to produce equilibrium. It is evident that these latter, as they produce the same effect as the body under the same circumstances, must have the same weight.

56. Sensibility of the Balance.—A balance is said to be more or less sensitive when the beam, supposed to be originally horizontal, is more or less inclined for a given difference of weights. The sensibility depends, in the first place, on the friction of the axis against its supports. In carefully constructed balances this axis is formed by the edge of a triangular prism of very hard steel, called a knife-edge, which rests upon a plane of steel or agate. In this way, as rotation takes place about a very fine axis, and as, besides, the materials employed are very hard, the friction is rendered exceedingly small.

Supposing friction to be eliminated, the sensibility of the balance depends upon the weight of the beam, its length, and the distance between its centre of gravity and the axis of suspension. We shall proceed to investigate the influence of these different elements.

Let A and B be the points from which the scale-pans are suspended, O the axis about which the beam turns, and G the centre of gravity of the beam. If when the scale-pans are loaded with equal weights, we put into one of them an excess of weight p, the beam will become inclined, and will take a position such as A'B', turning through an angle which we will call α, and which is easily calculated.

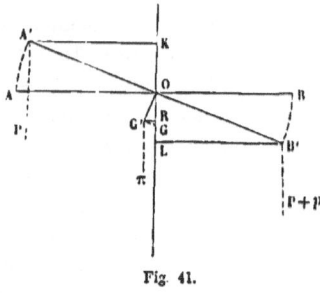

Fig. 41.

In fact, let the two forces P and P + p act at A' and B' respectively, where P denotes the less of the two weights, including the weight of the pan. Then the two forces P destroy each other in consequence of the re-sistance of the axis O; there is left only the force p applied at B', and the weight π of the beam applied at G', the new position of the centre of gravity. These two forces are parallel, and are in equilibrium about the axis O, that is, their resultant passes through the point O. The distances of the points of application of the forces from a vertical through O are therefore inversely proportional to the forces themselves, which gives the relation

$$\pi.\ \text{G'R} = p.\ \text{B'L}.$$

But if we call half the length of the beam l, and the distance OG r, we have

$$\text{G'R} = r \sin \alpha, \quad \text{B'L} = l \cos \alpha,$$

whence $\pi r \sin \alpha = pl \cos \alpha$, and consequently

$$\tan \alpha = \frac{pl}{\pi r}. \qquad (a)$$

The formula (a) contains the entire theory of the sensibility of the balance when properly constructed. We see, in the first place, that tan α increases with the excess of weight p, which was evident beforehand. We see also that the sensibility increases as l increases and as π diminishes, or, in other words, as the beam becomes longer and lighter. At the same time it is obviously desirable that, under the action of the weights employed, the beam should be stiff enough to undergo no sensible change of shape. The problem of the balance then consists in constructing a beam of the greatest possible length and lightness, which shall be capable of supporting the action of given forces without bending.

Fortin, whose balances are justly esteemed, employed for his beams bars of steel placed edgewise; he thus obtained great rigidity, but certainly not all the lightness possible. At present the makers of balances employ in preference beams of copper or steel made in the form of a frame, as shown in Fig. 42. They generally give them the shape of a very elongated lozenge, the sides of which are connected by bars variously arranged. The determination of the best shape is, in fact, a special problem, and is an application on a small scale of that principle of applied mechanics which teaches us that hollow pieces have greater resisting power in proportion to their weight than solid pieces, and consequently, for equal resisting power, the former are lighter than the latter. Aluminium, which with a rigidity nearly equal to that of copper, has less than one-fourth of its density, seems naturally marked out as adapted to the construction of beams. It has as yet, however, been little used.

The formula (a) shows us, in the second place, that the sensibility increases as r diminishes; that is, as the centre of gravity approaches the centre of suspension. These two points, however, must not coincide, for in that case for any excess of weight, however small, the beam would deviate from the horizontal as far as the mechanism would permit, and would afford no indication of approach to equality in the weights. With equal weights it would remain in equilibrium in any position. In virtue of possessing this last property, such a balance is called *indifferent*. Practically the distance between the centre of gravity and the point of suspension must not be less than a certain amount depending on the use for which the balance is designed. The proper distance is determined by observing what difference of weights corresponds to a division of the graduated arc along which the needle moves. If, for example, there are 20 divisions on each side of zero, and if 2 milligrammes are necessary for the total displacement of the needle, each division will correspond to an excess of weight of $\frac{2}{20}$ or $\frac{1}{10}$ of a milligramme. That this may be the case we must evidently have a suitable value of r, and the maker is enabled to regulate this value with precision by means of the screw which is shown in the figure above the beam, and which enables him slightly to vary the position of the centre of gravity.

In the above analysis we have supposed that the three points of suspension of the beam and of the two scale-pans are in one straight line; in which case the value of tan a does not include P, that is, the sensibility is independent of the weight in the pans. This follows from the fact that the resultant of the two forces P passes through O, and is thus destroyed, because the axis is fixed. This would not be the case if, for example, the points of suspension of the pans were above that of the beam; in this case the point of application of the common load is above the point O, and, when the beam is inclined, acts in the same direction as the excess of weight; whence the sensibility increases with the load up to a certain limit, beyond which the equilibrium becomes unstable.[1] On the other hand, when the points of suspension of the pans are below that of the beam, the sensibility increases as the load diminishes, and, as the centre of gravity of the beam may in this case be above the axis, equilibrium may become unstable when the load is less than a certain amount. This variation of the sensibility with the load is a serious disadvantage; for, as we have just shown, the displacement of the needle is used as the means of estimating weights, and for this purpose we must have the same displacement corresponding to the same excess of weight. If we wish to employ either of the two above arrangements, we should weigh with a constant load. The method of doing so, which constitutes a kind of double weighing, consists in retaining in one of the pans a weight equal to the maximum load. In the other pan is placed the same load subdivided into a number of marked weights. When the body to be weighed is placed in this latter pan, we must, in order to maintain equilibrium, remove a certain number of weights, which evidently represent the weight of the body.

We may also remark that, strictly speaking, the sensibility always depends upon the load, which necessarily produces a variation in the friction of the axis of suspension. Besides, it follows from the nature of bodies that there is no system that does not yield somewhat even to the most feeble action. For these reasons, there is a decided advantage in operating with constant load.

57. Suspension of the Scale-pans.—A fundamental condition of the correctness of the balance is, that the weight of each pan and of the load which it contains should always act exactly at the same point, and therefore at the same distance from the axis of suspension. This important result is attained by different methods. The arrange-

---

[1] This is an illustration of the general principle, applicable to a great variety of philosophical apparatus, that a maximum of sensibility involves a minimum of stability; that is, a very near approach to instability. This near approach is usually indicated by excessive slowness in the oscillations which take place about the position of equilibrium.

ment represented in Fig. 42 is one of the most effectual. At the extremities of the beam are two knife-edges, parallel to the axis of rotation, and facing upwards. On these knife-edges rests, by a hard plane surface of agate or steel, a stirrup, the front of which has been taken away in the figure. On the lower part of the stirrup rests another knife-edge, at right angles to the former, the

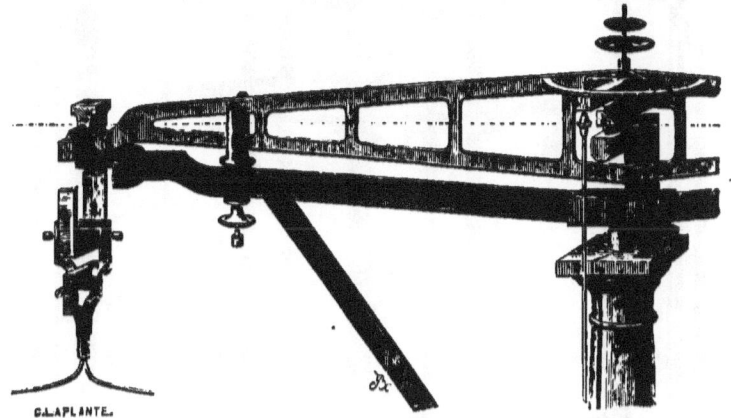

G.LAPLANTE.

Fig. 42.—Beam of Balance.

two being together equivalent to a universal joint supporting the scale-pan and its contents. By this arrangement, whatever may be the position of the weights, their action is always reduced to a vertical force acting on the upper knife-edge.

Fig. 43 represents a balance of great delicacy, with the glass case that contains it. At the bottom is seen the extremity of a lever, which enables us to raise the beam, and thus avoid wearing the knife-edge when not in use. At the top may be remarked an arrangement employed by some makers, consisting of a horizontal graduated circle, on which a small metallic index can be made to travel; its different displacements, whose value can be determined once for all, are used for the final adjustment to produce exact equilibrium.

58. Densities.—If we weigh equal volumes of different bodies in nature, we find that they have different weights. Thus, a litre of water weighs 1 kilogramme, a litre of mercury weighs 13·6 kilos., a litre of alcohol 0·79 kilos. This we express by saying that different bodies have different densities. It is evidently important to

know the density of the different substances with which we have to
deal; for this is, in fact, a fundamental element of their physical
constitution. Attempts have therefore been made to form a list
containing the weight of a given volume of each of the substances

BONNAFOUX   C. LAPLANTE. SC.

Fig. 48.—Balance for Purposes of Accuracy.

known in nature. The simplest way of determining the density
of a substance is to weigh a certain known volume of it, and to
divide the weight obtained by the volume; we shall thus obtain the
weight of unit volume.

The same object is attained indirectly by determining the *specific
gravity* of the substance; that is to say, the ratio of its density to
that of some standard substance whose density is known. The
standard substance commonly employed for this comparison is
distilled water at the temperature of maximum density (about 39°·1
Fahrenheit). The weight of a cubic foot of such water is 62·425
lbs. avoirdupois; hence the specific gravity of a substance multiplied

by 62·425 is the weight of a cubic foot of the substance in lbs., which may be called the *density of the substance in lbs. per cubic foot*.

In the metrical system the conversion of specific gravities into densities is much simpler; for since the gramme, kilogramme, and tonne are the weights respectively of a cubic centimetre, cubic decimetre, and cubic metre of distilled water at the temperature of maximum density, it follows that the same number which denotes the specific gravity of a substance, also denotes the weight of a cubic centimetre of the substance in grammes, the weight of a cubic decimetre in kilogrammes, or the weight of a cubic metre in tonnes. In other words, the specific gravity is equal to the density, whether the latter be expressed in grammes per cubic centimetre, in kilogrammes per cubic decimetre, or in tonnes per cubic metre.

If V denote the volume of a body, P its weight, and D its density, or weight per unit volume, we have

$$D = \frac{P}{V}, \qquad P = VD, \qquad V = \frac{P}{D};$$

so that if any two of these three elements are given, the third can be computed.

EXAMPLE I.—What is the weight of a mass of granite of 84 cubic metres, the density of granite being 2·75? The formula gives P = 84 × 2·75 = 231 tonnes.

EXAMPLE II.—What is the volume of 1000 kilos. of mercury, the density of mercury being 13·6?

$V = \frac{1000}{13·6} = 73·5$ litres, since the litre is equal to the cubic decimetre.

59. Experimental Determination of Densities.—The following is one of the simplest methods for the practical determination of densities. We begin by weighing the body. Suppose, for example, its weight to be 10 grammes. It is then placed upon one of the scale-pans of a balance, along with a flask with a wide neck, of a form such as is shown in Fig. 44, and exactly full of water; these are balanced in the other scale by weights of any kind. The body is then introduced into the flask, and evidently displaces a volume of water equal to its own volume. If now we close the flask, taking care that it is filled to the same level as before, wipe it, and put it back on the scale-pan, there will not be equilibrium. In order to re-establish equilibrium we must add, suppose, 2·5 grammes; this is the weight of a volume of water equal to that of the body; the specific gravity of the latter then is $\frac{10}{2\cdot5} = 4$. When we wish to determine the density of a liquid, we employ a flask (Fig. 45), the upper extremity of which terminates in a narrow tube on which there is a mark. After

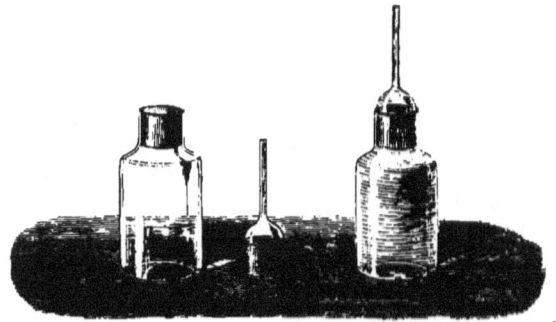

Fig. 44.—Specific Gravity Flask for Solids.

Fig. 45.—Specific Gravity Flask for Liquids.

having weighed the flask empty, we fill it successively with the liquid and with water, as far as the mark. This gives us the weights of equal volumes of water and of the liquid whose density is to be determined; and the quotient obtained by dividing the former by the latter is the specific gravity of the liquid.

There is generally some difficulty in filling the flask, on account of the very small diameter of the tube. The usual way is to put a little liquid into the cup at the upper extremity of the tube, and press it with the stopper; this pressure is generally sufficient to make the liquid pass into the flask. These two methods for determining densities are susceptible of great precision.

Other methods will be described in Chapter x.

In the following table we give the specific gravities of some liquids and solids.

Liquids, at Temperature of Freezing Water.

Water, sea, ordinary,	1·026	Oil, linseed,	·940
Alcohol, pure,	·791	,, olive,	·915
,, proof spirit,	·916	,, whale,	·923
Ether,	·716	,, of turpentine,	·870
Mercury,	13·596	Blood, human,	1·055
Naphtha,	·848	Milk, of cow,	1·03

SOLIDS.

Brass, cast, ... 7·8 to 8·4	Ice, ... ·92
,, wire, ... 8·54	Basalt, ... 3·00
Bronze, ... 8·4	Brick, ... 2 to 2·17
Copper, cast, ... 8·6	Brickwork, ... 1·8
,, sheet, ... 8·8	Chalk, ... 1·8 to 2·8
,, hammered, ... 8·9	Clay, ... 1·92
Gold, ... 19 to 19·6	Glass, crown, ... 2·5
Iron, cast, ... 6·95 to 7·3	,, flint, ... 3·0
,, wrought, ... 7·6 to 7·8	Quartz (rock-crystal), ... 2·65
Lead, ... 11·4	Sand, ... 1·42
Platinum, ... 21 to 22	Fir, spruce, ... ·48 to ·7
Silver, ... 10·5	Oak, European, ... ·69 to ·99
Steel, ... 7·8 to 7·9	Lignum-vitæ, ... ·65 to 1·33
Tin, ... 7·3 to 7·5	Sulphur, octahedral, ... 2·05
Zinc, ... 6·8 to 7·2	,, prismatic, ... 1·98

The unit of specific gravity is the specific gravity of pure water at the temperature of maximum density (39°·1 Fahr.)

The weight of a cubic foot of any substance is equal to 62·425 lbs. avoirdupois, multiplied by its specific gravity.

The weight of a cubic centimetre of any substance, in grammes, is equal to its specific gravity.

The weight of a litre (or cubic decimetre) of any substance, in kilogrammes, is equal to its specific gravity.

The weight of a gallon of any liquid, in lbs. avoirdupois, is equal to its specific gravity multiplied by 10.

CHAPTER VIII.

HYDROSTATICS.

60. Transmission of Pressure.—The peculiar constitution of liquids (§ 19) involves some important properties as regards pressure and the transmission of pressure. If we suppose that in a vessel A (Fig. 46), full of liquid, an opening is made at P, and a certain pressure applied there by means of a piston, the effect of this

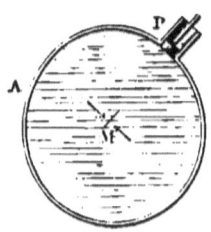

Fig. 46.—Transmission of Pressures.

pressure will be to bring the molecules closer together, and, consequently, to create a repulsive action between them. As this result takes place throughout the whole extent of the mass, it is evident that each point in the sides of the vessel will be pressed, and that thus the effect of the single pressure will be transmitted in an infinite number of different directions. This kind of irradiation of pressure in fluids is a distinctive characteristic, and has most important applications.

The pressure exerted at P takes effect not only upon the sides of the vessel, but also at every point in the liquid. Thus a small plane lamina, which we will suppose placed at M, will be subjected to two equal and opposite pressures upon its two faces. It is also very important to remark, that on account of the uniform nature of the liquid, these pressures will not change in magnitude if we suppose the lamina turned round so as to take different directions in the liquid mass, for there is evidently no reason why the pressure should be greater in one direction than in another.

61. Direction of Pressure.—The same reason of symmetry shows us that, at each of their points of application, these pressures are normal or perpendicular to the surface; for if any reason were assigned for their inclination in a certain direction, a similar reason could also be assigned for their inclination in any other direction. This important truth may also be inferred from observing that if at any point M in the side of a vessel (Fig. 47) the pressure PM was not normal, it could be decomposed into two: one, MN, along the normal to the surface, which would be destroyed by the resistance of the surface; the other, MA, along the surface itself, which latter force would cause a sliding motion in the liquid molecule at M, which serves as the medium of transmitting the pressure.

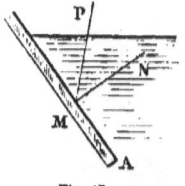

Fig. 47.

Experiment enables us, if not rigorously to demonstrate this principle, at least to show that the direction of transmitted pressure is sensibly normal. For example, if a sphere is taken, pierced with several holes, and containing a liquid, which is compressed by means of a piston in a tube communicating with the sphere, the liquid is seen to spout out in jets which take a curvilinear form under the action of gravity, but which at their origin appear perpendicular to the spherical surface. The effect is the more striking the greater the pressure exerted on the piston.

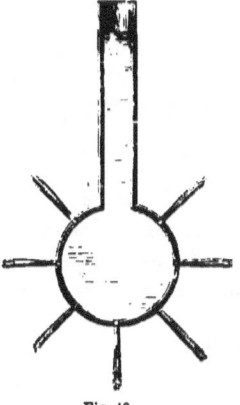

Fig. 48.

62. Pascal's Principle, or the Equal Transmission of Pressure in all Directions.—If we have a vessel A full of a liquid (Fig. 49), and if at a certain point P a certain pressure be exerted by means of a piston of the area of a square inch, suppose, each square inch of the sides of the vessel will be subjected to an equal pressure. If then at any point an opening is made of the area of a square inch, and closed with a piston, it will be necessary, in order to prevent the piston from moving, to apply to it from outside a pressure equal to that which is directly applied to the piston P. A lamina of the same area placed in any direction in the liquid will also be subjected to an equal pressure on each of its two faces.

Hence it follows that if we suppose a piston of the area of two square inches closing a corresponding opening, since each of these two inches of area receives a pressure equal to that which acts upon

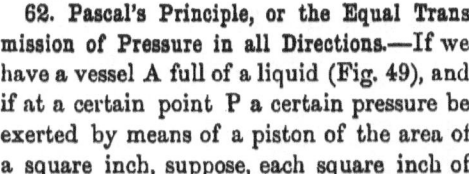

P, the whole piston will receive a double pressure; whence we see that in general the transmitted pressure should vary as the area of the surface pressed.

This is the form in which Pascal enunciated the principle in his

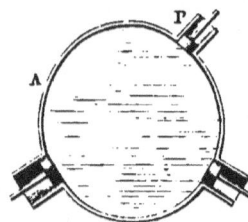

Fig. 49.—Pascal's Principle.

celebrated treatise on the *Equilibrium of Liquids.* "If a vessel full of water, closed on all sides, has two openings, the one a hundred times as large as the other, and if each be supplied with a piston which fits exactly, a man pushing the small piston will exert a force which will equilibrate that of a hundred men pushing the piston which is a hundred times as large, and will overcome that of ninety-nine. And whatever may be the proportion of these openings, if the forces applied to the pistons are to each other as the openings, they will be in equilibrium."

In general, let P be the pressure exerted upon a liquid by the aid of a piston of superficial extent S, each unit of surface of this piston will be subjected to a pressure $\frac{P}{S}$, and, as a consequence, on each unit of surface of the sides of the vessel a similar pressure will be produced. If then openings of areas S', S''.... are made at different points, and closed with pistons, we must, in order to prevent the pistons from moving, apply to them forces P', P''.... equal respectively to $S'\frac{P}{S}$, $S''\frac{P}{S}$.... which gives the following equations:—

$$P' = S'\frac{P}{S}, \quad P'' = S''\frac{P}{S}, \quad \text{or } \frac{P}{S} = \frac{P'}{S'} = \frac{P''}{S''}.$$

63. Pascal's principle leads to a consequence which we may verify by experiment. If into a system of two tubes in communication with each other, and of unequal sectional area, we introduce a liquid, it will stand at the same height in both branches. If then we place a piston on the liquid in the narrow tube, and subject it to a certain pressure P, this pressure will be transmitted to the liquid, which will be forced back into the large tube; to hinder this motion we must place a piston in the large tube, and apply to it a force which has the same ratio to the force P as the area of the larger piston to that of the smaller. If, for example, the former has an area 16 times that of the latter, a pressure of 1 pound exerted at one of the extremities of the liquid column will produce a pressure of 16 pounds at the other extremity. We thus see that a small force may be made to produce a very great one. This is the principle of the hydraulic press, a machine which we shall describe further on.

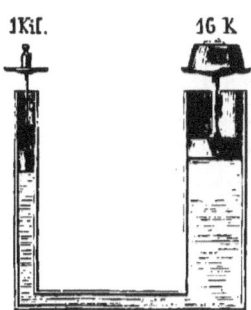

Fig. 50.—Principle of the Hydraulic Press.

We must remark, however, that if work is to be done, the one piston must displace the other; and it is very evident that, on account of the difference of section, if the small piston moves through a certain length, the large piston will move through one-sixteenth of that length; so that in this apparatus we have a direct verification of this general principle of mechanics, that *what is gained in force is lost in velocity.*

This twofold observation has been clearly enunciated by Pascal, who expresses himself in the following manner at the end of the passage which we have already quoted:—" Whence it appears that a vessel full of water is a new principle of mechanics, and a new machine for the multiplication of force to any required degree, since one man will by this means be able to raise any given weight.

"It is, besides, worthy of admiration that in this new machine we find that constant rule which is met with in all the old ones, such as the lever, wheel and axle, screw, &c., which is that the distance is increased in proportion to the force; for it is evident that as one of these openings is a hundred times as large as the other, if the man who pushes the small piston drives it forward one inch, he will drive the large piston backward only one-hundredth part of that length."

If we endeavoured to perform the preceding experiment in order to demonstrate experimentally the principle of Pascal, we should arrive at only an approximate verification; for to obtain an accurate experiment it would be necessary that the pistons should fit their openings with great exactness, and this would involve a large amount of friction.

The verification would be still more difficult if we endeavoured to perform the experiment described in § 62, for in this case, besides the cause of error which we have just mentioned, the phenomena would be complicated by the action of gravity, which of itself produces variable pressures on the different openings, according to their depth below the surface of the liquid.[1] In fact, the principle of Pascal is an abstract principle, a kind of general synthesis of phenomena which cannot be made the subject of direct demonstration. It is by the constant agreement of its consequences with our observation that the authority and legitimacy of the principle are established; we shall see from all that follows how complete and invariable is that agreement.

64. Fundamental Principle of Equilibrium in Heavy Liquids. Surfaces of Equal Pressure.—Pascal's principle is a general consequence of the constitution of liquids, and is independent of the action of gravity. By introducing this latter force we arrive at special results, which we shall describe in succession. The most important, which may be considered as the fundamental rule in hydrostatics, consists in the fact *that the different points in a horizontal layer of a heavy liquid*

Fig. 51.

are subject to the same pressure. Let us consider two points, A and B (Fig. 51), situated in the same horizontal plane in a heavy liquid. If we suppose that A and B are the centres of two small plane surfaces which are vertical and parallel, we may consider these surfaces as the bases of a very narrow horizontal cylinder of liquid. As this cylinder is in separate equilibrium in the general mass, we may conclude that its bases A and B are subject to equal and contrary pressures in the direction of the arrows shown in the figure; for the remaining pressures due to the surrounding liquid act in directions perpendicular to the axis of the cylinder, and thus cannot influence the equilibrium in the direction of the axis. The two elements A and B are therefore subject to the same pressure in one direction; but the pressure at a point is equal in all directions (§ 60); and as A and B are any points in the same horizontal layer, it follows that the pressures at all points in the same horizontal layer are equal. We may add, as a consequence of this, that the density is also the same at all points of a horizontal layer. On account of the slight compressibility of liquids, the variation of their density with the depth

---

[1] The pressures on the different pistons (supposed equal in area) would differ from each other by constant amounts, depending on their differences of level; but, on account of this constancy, any *increase of pressure* on one piston would require an *equal increase of pressure* on every other in order to the maintenance of equilibrium.

is scarcely sensible; but the above result is true for all heavy fluids, whether compressible or incompressible.

In proceeding from one surface of equal pressure to another, the pressure increases or diminishes according as the depth increases or diminishes. Thus, for example, if we consider a small horizontal element m of a horizontal layer AB, and conceive the vertical cylinder mm' reaching to the horizontal layer CD, it is quite clear that, independently of the pressure at m', which is transmitted undiminished to m, this latter element is subject to a pressure equal to the weight of the liquid contained in the cylinder mm'.

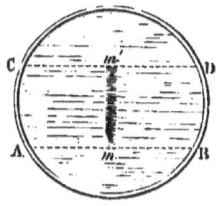

Fig. 52.

If we call s the area of the element m, h the distance between the two horizontal layers, and d the weight of unit volume of the liquid, the volume of the cylinder is expressed by sh, and its weight by shd. This last expression, therefore, represents the variation of pressure for an element of area s, when its depth below the surface varies by a quantity equal to h; and dividing by the area s we see that hd is the expression for the difference of pressure per unit area, corresponding to a difference of depth h.

65. Free Surface.—It follows from the foregoing principles that the free surface of a heavy liquid must be horizontal. We have already given an experimental demonstration of this important fact.

Fig. 53.

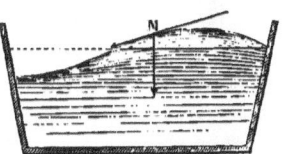

Fig. 54.

It could also have been predicted from *à priori* considerations. Let CD (Fig. 53) be the free surface, and m, m' two small equal elements of surface, in the horizontal layer AB. These two elements must be subject to equal pressures, which are evidently represented by the weights of the cylinders mn, $m'n'$; these cylinders must, therefore, have the same height, or the points n and n' must be in the same horizontal plane.

The same conclusion is arrived at by observing that if at any point whatever of the surface M (Fig. 54) the liquid did not assume a horizontal position, the weight of the liquid molecule at M could be decomposed into two forces, one perpendicular to and one along the surface of the liquid. The effect of the first would merely be to compress the liquid, and it would be destroyed by the reaction of the liquid; but the second would produce a displacement of the molecule. Equilibrium, therefore, can only exist on condition of the annihilation of this second component; that is, the surface must be horizontal at all points.

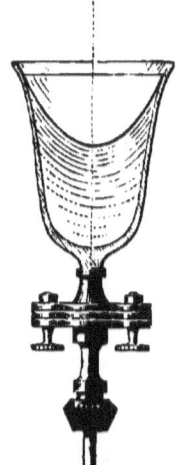

This mode of reasoning shows us that, in general, when the liquid mass is subjected to the action of any number of forces, it is necessary for equilibrium that the free surface be, at all points, perpendicular to the resultant of the acting forces. If, for instance, we place upon the whirling table a glass containing a liquid (Fig. 55), and give it a rotatory motion, we shall see the surface become hollowed, and assume a curvilinear form. In fact, each of the molecules is subjected simultaneously to the action of gravity and of centrifugal force; and it is the resultant of these two forces which must be everywhere perpendicular to the free surface. It is easily shown that this surface must be a paraboloid of revolution, so that the section represented in the figure is a parabola.

66. Pressure upon the Bottom of Vessels.—If we consider a heavy liquid placed in a vessel the bottom of which is formed by a *plane horizontal surface*, it is easy to determine the pressure exerted by the liquid upon this plane. Let ABMN (Fig. 56) be a vessel filled with liquid to the level MN, and *m* an element of surface on the bottom AB. On *m* suppose a small vertical cylinder to stand, meeting the horizontal layer LL' in *m'*. On the element *n*, equal to *m*, and in the horizontal layer LL', suppose a vertical cylinder to stand, cutting the horizontal layer RR' in *n'*. Suppose another similar vertical cylinder on *r*, an element equal to *m*, and let this cylinder meet SS' in *r'*; it is evident that, by continuing this construction, we shall finally arrive, whatever be the shape of the vessel, at a cylinder SS', which will extend upwards to the free surface

Fig. 55.—Rotating Vessel of Liquid.

MN. Now, the element m is subject to a pressure greater than that at m' by a quantity equal to the weight of the cylinder mm'. Simi-rally m' is subject to a pressure which is equal to that at n, and greater than that at n' by a quantity equal to the weight of the cylinder nn'; whence it is evident, by pursuing this reason-ing, that the element m supports a pressure equal to the sum of the weights of the cylinders mm', nn', rr', ss'; that is, to the weight of a cylinder of liquid standing on the base m, and extending vertically upwards to the free surface. As all points in the bottom AB are subject to

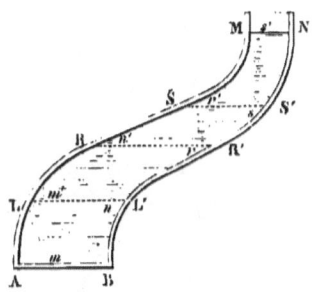

Fig. 56.—Pressure on the Bottom of Vessels.

the same pressure, it follows that the entire pressure on the bottom of the vessel is equal to the weight of a liquid column whose base is the bottom of the vessel, and height the vertical distance between the bottom and the free surface, or what is called the height of the liquid in the vessel.[1]

Let B be the area of the bottom of the vessel, H the height of the liquid, and D its weight per unit volume, then the pressure is expressed by the formula BHD.

If, for instance, in a vessel whose bottom is two square decimetres in area, there is a column of mercury of the height of $5\frac{1}{2}$ decimetres, the volume of this column, which measures the pressure, is $2 \times 5\frac{1}{2}$ = 11 cubic decimetres, and its weight[2] is $11 \times 13\cdot59 = 149\cdot49$ kilo-grammes.

67. Experiment of Pascal's Vases.—The preceding proposition shows that the pressure on the bottom of a vessel depends only upon the area of the bottom and the height of the liquid, the form of the vessel being quite immaterial. In order to verify this fact, Pascal, contrived an experiment which, with some modifications, is now commonly introduced in courses of physics. The apparatus employed is a tripod (Fig. 57) supporting a ring, into which can successively

[1] In these remarks we neglect the atmospheric pressure which is exerted upon the free surface and transmitted to the bottom of the vessel.

[2] This example illustrates the convenience of the metrical system. The weight of a cubic decim. of water is 1 kilo., and the sp. gr. of mercury is 13·59; hence the weight of a cubic decim. of mercury is 13·59 kilos.

be screwed three vessels of different shapes, one widened upwards, another cylindrical, and the third tapering upwards. At the lower part of the ring is a disc, supported by a thread fixed to one of the scales of a balance. Weights placed in the other scale keep the disc pressed against the ring with a certain force. Let the cylindrical vase be placed upon the tripod, and filled with water until the

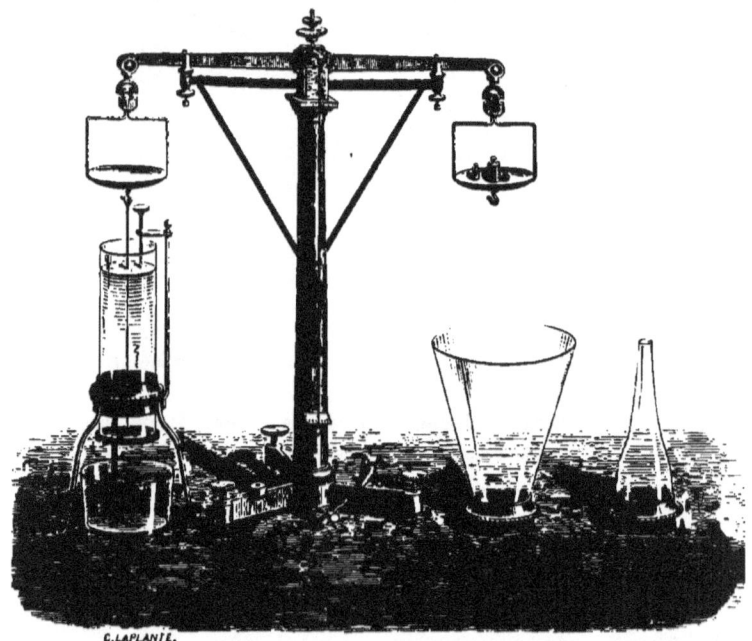

C. LAPLANTE.

Fig. 57.—Experiment of Pascal's Vases.

pressure exerted on the disc detaches it from the ring. An indicator marks the level of the liquid when this takes place. Let the experiment be repeated with the two other vases, and the disc will be detached when the water has reached the same height; showing plainly that the pressure on the bottom of a vessel is independent of the shape of the vessel.

But we may go further; for in the case of the cylindrical vessel it is evident that the pressure on the bottom is equal to the weight of the contained liquid. Now this weight is necessarily equal to that which counterpoised it in the other scale of the balance; hence in all three cases the pressure on the bottom of the vessel is equal to the weight of a liquid column with the bottom as base and of the same height as the liquid in the vessel.

68. Upward Pressure.—The pressure exerted at any point of a liquid mass being the same in all directions, a horizontal surface facing downwards should be subjected to an upward pressure equal to the downward pressure which would be exerted if the liquid were acting in the other direction. Suppose we take a tube open at both ends (Fig. 58), and apply to the lower end a flat cover. If we plunge the tube into a liquid, the liquid will press up the cover against the bottom of the tube with a force which increases as the tube is plunged deeper in the liquid. If we then pour liquid into the tube, this will produce a downward pressure upon the cover, and when the level of the liquid is the same inside the tube as outside the cover will be

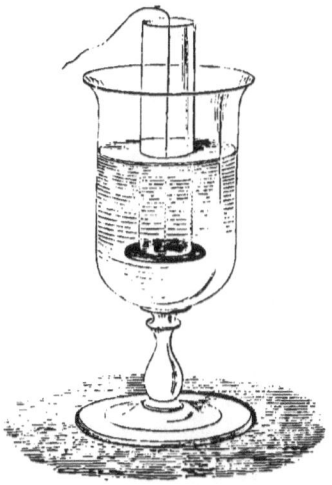

Fig. 58.—Upward Pressure.

detached, the upward pressure being destroyed by the pressure exerted by the liquid inside the tube in the contrary direction.

69. Total Pressure. Resultant Pressure on Vessel. Hydrostatic Paradox. —When a vessel of any shape is filled with a liquid, normal pressures are exerted against all points of its sides, increasing with the depth, and equal in each case to the pressure in the corresponding horizontal layer of the liquid. We may suppose a summation of all the pressures thus exerted upon the different superficial elements of the sides; this gives what is called the *total pressure* exerted by the liquid (see § 72).

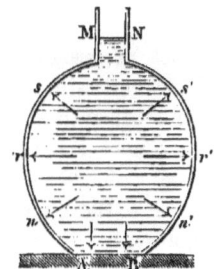

Fig. 59.—Total Pressure.

This total pressure must not be confounded with the resultant pressure which is transmitted to the stand on which the vessel rests. In fact we see that, of the elements of pressure, some are transmitted *entire* to the stand; these are the vertical pressures exerted upon the bottom AB; others, those for instance at n and n', are only transmitted in part, because their direction is oblique; the horizontal pressures at r and r' are evidently without influence; and the pressures exerted at s and s' tend to raise the vessel. It is by the combination and *composition* of these pressures of different intensity and direction that the resultant pressure upon the stand on which the vessel rests is obtained.

Hydrostatic Paradox.—It has been thought paradoxical that vessels whose bottoms were subjected to equal pressures did not transmit equal pressures to the stand on which they were placed. In reality, nothing is less paradoxical; the pressure on the bottom of the vessel is only one of the elements which combine to produce the resultant pressure transmitted to the stand.

70. Composition of Pressures.—It may be regarded as quite evident that this latter pressure is in all cases equal to the weight of the liquid; which amounts to saying that if we place a vessel containing liquid in one scale of a balance, we must, in order to equilibrate it, place in the other scale weights equal to the sum of the weights of the liquid and the vessel. This result is also easily shown from *à priori* considerations in some simple cases.

Thus in the case of a cylindrical vessel ABDC (Fig. 60) it is evi-

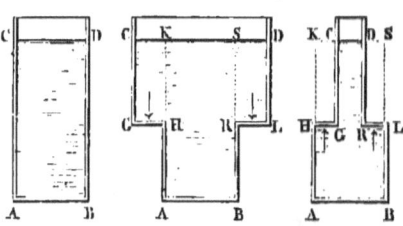

Fig. 60.—Hydrostatic Paradox.

dent that the only pressure transmitted to the stand is that exerted upon the bottom, which is equal to the weight of the liquid. In the case of the vessel which is wider at the top, the stand is subjected to the weight of the liquid column ABSK, which presses on the bottom AB, together with the columns GHKC, RLDS, pressing on GH and RL; the sum of which weights composes the total weight of liquid contained in the vessel. Finally, in the third case, the pressure on the bottom AB, which is equal to the weight of a liquid column ABSK, must be diminished by the pressures in the opposite direction on HG and RL. These last being represented by liquid columns HGCK, RLSD, there is only left to be transmitted to the stand a pressure equal to the weight of the water in the vessel. By the application of the ordinary rules for the composition of forces, it can easily be shown that this result is perfectly general. Here, then, we have Pascal's principle leading to a conclusion which is obviously true, and which therefore furnishes a confirmation of the principle itself.

71. Motion produced by the Flowing of a Liquid.—The demonstration to which we have just referred amounts to showing, by the analysis of the different pressures, that the horizontal components of these pressures equilibrate each other, and that the vertical components are equivalent to a single force equal to the weight of the liquid. The evidence of experiment is as strong for the first part of this proposition as for the second. If we place a vessel (Fig. 61) in such a position as to be very susceptible of motion in a hori-

Fig. 61.—Backward Movement of Discharging Vessel.

zontal direction, whether by suspending it by a thread, or by floating it in water, and under these circumstances fill it with liquid, however

great may be the mobility of the apparatus, not the slightest displacement is observed. This proves that the horizontal components of pressure equilibrate one another. This equilibrium is effected through the medium of the vessel; but if we suppose an opening made at any point of the vessel, the liquid will run out, and the pressure exerted at the point diametrical-

Fig. 62.—Hydraulic Tourniquet.

ly opposite will propel the vessel in the opposite direction to that in which the liquid is escaping.

This observation explains the motion of the apparatus called the *hydraulic tourniquet.* It consists (Fig. 62) of a vessel capable of rotation about a vertical axis, having at its lower extremity a tube, the two portions of which are bent in opposite directions, and left open at the ends to allow the liquid to escape. The reaction at the points opposite to the openings causes the rotatory motion of the apparatus.

When the velocity of efflux is sufficiently great, the motion can be employed for practical purposes, and hydraulic engines based on this principle have frequently been proposed and tried; Barker's mill is one of the best known.

72. Centre of Pressure.—When we consider the particular case of the pressure exerted by a liquid upon a *plane* surface, the different elements of pressure being all parallel, it may be required to determine the point of application of the resultant pressure. This point is called the *centre of pressure.* The centre of pressure does not coincide with the centre of gravity; it is always situated below this latter, since the elementary forces which must be combined in order to obtain it, instead of being uniformly distributed over the surface, increase with the depth.

The investigation of the centre of pressure constitutes a separate chapter of mathematical physics, and we shall not enter upon it here; we shall simply examine a particular case, fitted to give accurate ideas with regard to this point.

Let a rectangular surface RB be immersed in a liquid, which extends

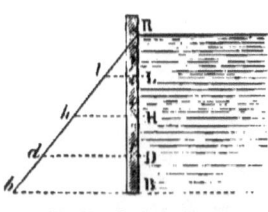

Fig. 63.—Centre of Pressure.

as far as R; we may suppose it, for instance, a flood-gate or the side of a dam for holding water. The pressure goes on increasing from the point R, where it is zero, to the point B, where it attains its maximum value; it has the same value for all points in the same horizontal line, and is at each point proportional to the distance from the surface of the water.

If, then, at the point B we draw a perpendicular B*b* equal to RB, and join R*b*, the different parallel lines D*d*, H*h*, L*l*, in the triangle RB*b* will be proportional to the pressures at the points D, H, L. The composition of these pressures, then, amounts to finding the centre of gravity of the triangle R*b*B; but the height of this from the base is one-third of the height of the triangle; the centre of pressure thus lies at one-third of the height RB. It is further evident by symmetry that it will lie on the line joining the middle points of the upper and lower sides of the rectangle.

As for the total pressure on RB, it may be obtained in the special case under consideration by observing, that since the pressure increases uniformly from R to B, its average intensity is equal to the pressure at the middle point. The total pressure is therefore the same in amount (though not in distribution) as if the surface were pressed at all points by a body of water of half the height of RB.

Suppose the height of RB = 3 metres, and its breadth = 5 metres, the total pressure will be equal to the weight of $5 \times 1\cdot5 = 7\cdot5$ cubic metres of water, that is, to 7500 kilos., and will have for its resultant a single force of this amount applied not at the centre of gravity, but at the centre of pressure.

We may remark that the middle point of the height of the rectangle exactly corresponds to the centre of gravity of the figure, and it may be demonstrated in general that *the total pressure on any surface, whether plane or curved, is equal to the weight of a liquid column having that surface for base, and for height the distance of the centre of gravity from the surface of the water.*

CHAPTER IX.

73. Pressure of Liquids on Bodies immersed.—When a body is immersed in a liquid, the different points of its surface are subjected to pressures which obey the rules laid down in the preceding chapter. As these pressures increase with the depth, it is evident that those which tend to raise the body overcome those which tend to sink it, so that the resultant effect is a force in the direction opposite to that of gravity.

By means of an analysis, similar to that in § 70, it may be shown that this resultant upward force is exactly equal to the weight of the liquid displaced by the body.

This conclusion can very readily be verified in some simple cases: suppose, for example (Fig. 64), a right cylinder plunged vertically in a liquid, and let us examine the effect of the different pressures exerted by the liquid upon its surface. It is evident, in the first place, that if we consider any point on the sides of the cylinder, the normal and horizontal pressure on that point is destroyed by the equal and contrary pressure at the point diametrically opposite; and, as the same is the case for all similar points, we see that the horizontal pressures destroy each other. As regards the vertical pressures on the ends, one of them, that on the upper end AB, is in a downward direction, and equal to the weight of the liquid column ABNN; the other, that on the lower end CD, is in an upward direction, and equal to the weight of the liquid column CNND; this latter pressure exceeds the former by the weight of the liquid cylinder ABCD, so that the resultant effect of the pressure is to raise the body with a force equal to the weight of the liquid displaced.

By a synthetic process of reasoning, we may, without having recourse to the analysis of the different pressures, show that this conclusion is perfectly general. Suppose we have a liquid mass in equilibrium, and that we consider specially the portion M (Fig. 65); this portion is likewise in equilibrium. If we suppose it to become solid, without any change in its weight or volume, equilibrium will

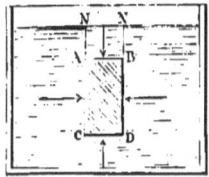
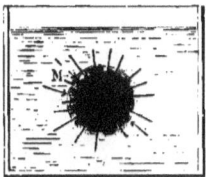

Fig. 64.   Principle of Archimedes.   Fig. 65.

still subsist. Now this is a heavy mass, and as it does not fall, we must conclude that the effect of the pressures on its surface is to produce a resultant upward pressure exactly equal to its weight, and acting in a line which passes through its centre of gravity. If we now suppose M replaced by a body exactly occupying its place, the exterior pressures remaining the same their resultant effect will also be the same.

The name *centre of buoyancy*, or *centre of displacement*, is given to the centre of gravity of the liquid displaced by a body immersed, and we see that we may always suppose that it is in this point that the upward pressure of the liquid is applied. The results of the above explanations may thus be included in the following proposition: *Every body immersed in a liquid is subjected to an upward vertical pressure equal to the weight of the liquid displaced, and applied at the centre of displacement.*

This proposition constitutes the celebrated principle of Archimedes. It is often enunciated in the following form: *Every body immersed in a liquid loses a portion of its weight equal to the weight of the liquid displaced.* This enunciation, though perhaps less correct than the former, is fundamentally identical with it; for if we weigh a body immersed in a liquid, the weight will evidently be diminished by a quantity equal to the upward pressure.

74. Experimental Demonstration of the Principle of Archimedes.—The following experimental demonstration of the principle of Archimedes is commonly exhibited in courses of physics:—

From one of the scales of a hydrostatic balance is suspended a hollow cylinder of copper, and below this a solid cylinder, whose volume is equal to the interior volume of the hollow cylinder; these are balanced by weights in the other scale. A vessel of water is then placed below the cylinders, in such a position that the lower cylinder shall be immersed in it. The equilibrium is immediately destroyed, and the upward pressure of the water causes the scale with the weights to descend. If we now pour water into the hollow cylinder, equilibrium will gradually be re-established; and the beam

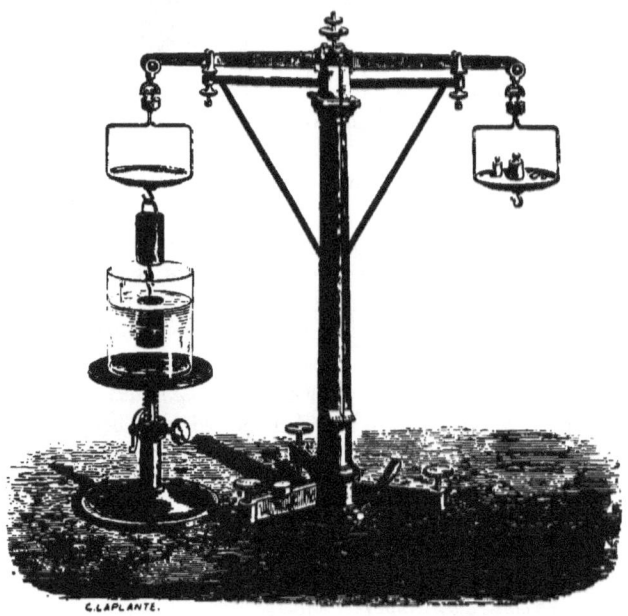

C. LAPLANTE.

Fig. 66.—Experimental Verification of Principle of Archimedes.

will be observed to resume its horizontal position when the hollow cylinder is full of water, the other cylinder being at the same time completely immersed. The upward pressure upon this latter is thus equal to the weight of the water added, that is, to the weight of the liquid displaced.

75. Body immersed in a Liquid.—It follows from the principle of Archimedes that when a body is immersed in a liquid, it is subjected to two forces: one equal to its weight and applied at its centre of gravity, tending to make the body descend; the other equal to the weight of the displaced liquid, applied at the centre of buoyancy, and tending to make it rise. There are thus three different cases to be considered:

(1.) The weight of the body may exceed the weight of the liquid displaced, or, in other words, the mean density of the body may be greater than that of the liquid; in this case, the body sinks in the liquid, as, for instance, a piece of lead dropped into water.

(2.) The weight of the body may be less than that of the liquid displaced; in this case the body rises partly out of the liquid, until the weight of the liquid displaced is equal to its own weight. This is what happens, for instance, if we immerse a piece of cork in water and leave it to itself.

(3.) The weight of the body may be equal to the weight of the liquid displaced; in this case, the two opposite forces being equal, the body takes a suitable position (§ 77) and remains in equilibrium.

These three cases are exemplified in the three following experiments (Fig. 67):—

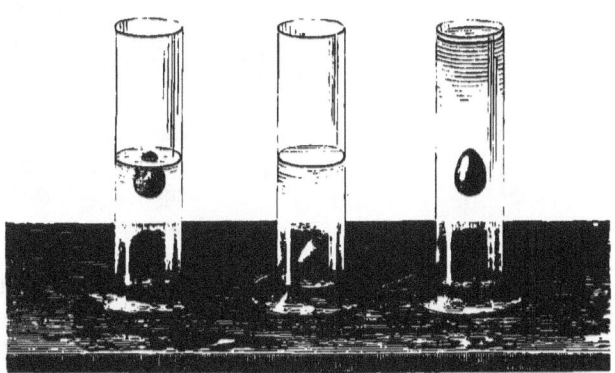

Fig. 67.—Egg Plunged in Fresh and Salt Water.

(1.) An egg is placed in a vessel of water; it sinks to the bottom of the vessel, its mean density being a little greater than that of the liquid.

(2.) Instead of fresh water, salt water is employed; the egg floats at the surface of the liquid, which is a little denser than it.

(3.) Fresh water is carefully poured on the salt water; a mixture of the two liquids takes place where they are in contact; and if the egg is put in the upper part, it will be seen to descend, and, after a few oscillations, remain at rest in a layer of liquid of which it displaces a volume whose weight is equal to its own. We may remark that, in this position the egg is in stable equilibrium; for, if it rises, the upward pressure diminishing, its weight tends to make it descend again; if, on the contrary, it sinks, the pressure increases and tends to make it reascend.

76. Cartesian Diver.—The experiment of the *Cartesian diver*, which is described in old treatises on physics, shows each of the different cases that can present themselves when a body is immersed. The diver (Fig. 68) consists of a hollow ball, at the bottom of which is a small opening O; a little porcelain figure is attached to the ball, and the whole floats upon water contained in a glass vessel, the mouth of which is closed by a strip of caoutchouc or a bladder. If we press with the hand on the bladder, the air is compressed, and the pressure, transmitted through the different horizontal layers, condenses the air in the ball, and causes the entrance of a portion of the liquid by the opening O; the floating body becomes heavier, and in

Fig. 68.—Cartesian Diver.

consequence of this increase of weight the diver descends. When we cease to press upon the bladder, the pressure becomes what it was before, some water flows out and the diver ascends. It must be observed, however, that as the diver continues to descend more and more water enters the ball, in consequence of the increase of pressure, so that if the depth of the water exceeded a certain limit, the diver would not be able to rise again from the bottom.

If we suppose that at a certain moment the weight of the diver becomes exactly equal to the weight of an equal volume of the liquid, there will be equilibrium; but, unlike the equilibrium in the experiment in § 75, this will evidently be *unstable*, for a slight movement either upwards or downwards will alter the resultant force so as to produce further movement in the same direction.

77. Relative Positions of the Centre of Gravity and Centre of Buoyancy. —In order that a floating body, wholly or partially immersed in a liquid, may be in equilibrium, it is evidently necessary that its weight be equal to the weight of the liquid displaced.

This condition, which is absolutely necessary, is, however, not sufficient; we require, in addition, that the action of the upward pressure should be exactly opposite to that of the weight; that is, that the centre of gravity and the centre of buoyancy be in the same vertical line; for if this were not the case, the two contrary forces would compose a couple, the effect of which would evidently be to cause the body to turn.

In the case of a body completely immersed, it is further necessary for stable equilibrium that *the centre of gravity should be below* the centre of buoyancy; in fact we see, by Fig. 69, that in any other

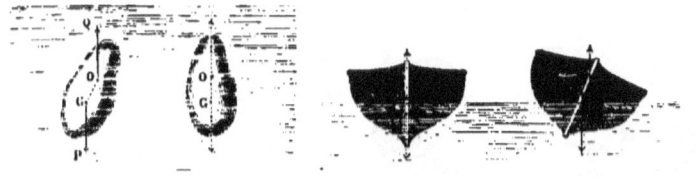

Fig. 69. Fig. 70.

Relative Positions of Centre of Gravity and Centre of Pressure.

position than that of equilibrium, the effect of the two forces applied at the two points G and O would be to turn the body, so as to bring the centre of gravity lower. But this is not the case when the body is only partially immersed, as most frequently happens. In this case it may indeed happen that, with stable equilibrium, the centre of gravity is below the centre of pressure; but this is not necessary, and in the majority of instances is not the case. Let Fig. 70 represent the lower part of a floating body—a boat, for instance. The centre of pressure is at O, the centre of gravity at G, considerably above; if the body is displaced, and takes the position shown in the figure, it will be seen that the effect of the two forces acting at O and at G is to restore the body to its former position. This difference from what takes place when the body is completely immersed, depends upon the fact that, in the case of the floating body, the figure of the liquid displaced changes with the motions of the body, and the centre of buoyancy moves towards the side on which the body is more deeply immersed. It will depend upon the form of the body whether this lateral movement of the centre of buoyancy is sufficient to carry it beyond the vertical through the centre of gravity. The two equal forces which act on the body will evidently turn it to or from the original position of equilibrium, according as the new centre of buoyancy lies beyond or falls short of this vertical.[1]

78. Advantage of Lowering the Centre of Gravity.—Although stable equilibrium may subsist with the centre of gravity above the centre of buoyancy, yet for a body of given form the stability is always increased by lowering the centre of gravity; as we thus lengthen the arm of the couple which tends to right the body when displaced. It is on this principle that the use of ballast depends.

79. Phenomena in apparent Contradiction to the Principle of Archi-

Fig 71.—Steel Needles Floating on Water.

medes.—A body cannot float in a liquid unless it have a density less than that of the liquid. This natural consequence of the principle of Archimedes seems at first sight to be contradicted by some well-known facts. Thus, for instance, if small needles are placed carefully on the surface of water, they will remain there in equilibrium (Fig. 71). It is on a similar principle that several insects *walk* on water (Fig. 72), that a great number of bodies of various natures, provided they be *very minute*, can, if we may so say, be placed on the surface of a liquid without penetrating into its interior. These curious facts depend on the circumstance that the small bodies in

---

[1] If a vertical through the new centre of buoyancy be drawn upwards to meet that line in the body which in the position of equilibrium was a vertical through the centre of gravity, the point of intersection is called the *metacentre*. Evidently when the forces tend to restore the body to the position of equilibrium, the metacentre is above the centre of gravity; when they tend to increase the displacement, it is below. In ships the distance between these two points is usually nearly the same for all amounts of heeling, and this distance is a measure of the stability of the ship.

We have defined the metacentre as the intersection of two lines. When these lines lie

question are not wetted by the liquid, and hence, in virtue of principles which will be explained in connection with capillarity (Chap.
xi.), depressions are formed around
them on the liquid surface, as represented in Fig. 73. The curvature of the liquid surface in the
neighbourhood of the body is very
distinctly shown by observing the
shadow cast by the floating body,

Fig. 72.—Insect Walking on Water.

when it is illumined by the sun; it is seen to be bordered by lumi-
nous bands, which are owing to the refraction of the rays of light in
the portion of the liquid bounded by a curvilinear surface.

The existence of the depression about the floating body enables us
to bring the condition of equilibrium in this special case under the
general enunciation of the principle of Archimedes.
Let M be a section of the body, CD the distance
to which the depression extends, and AB the corresponding portion of any horizontal layer; since
the pressure at each of the points of AB must be
the same as in the other parts of the layer, the

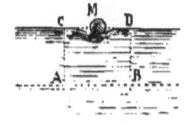

Fig. 73.

liquid acts in exactly the same way as if M did not exist, and the
cavity were filled by the liquid itself.

We may thus say in this case also that the weight of the floating
body is equal to the weight of the *liquid displaced*, understanding
by these words the liquid which would occupy the whole extent of
the depression due to the presence of the body.

80. Liquids in Superposition.—When liquids of different densities,
which do not readily mix, are placed in the same vessel, the particles
of the denser liquids unite and fall to the bottom, just as a solid body
sinks in a liquid of less density; finally, the liquids arrange themselves in the order of their respective densities, the surfaces of separation being horizontal. This fact is verified by means of the phial
called the *phial of the four elements*. It is a flask (Fig. 74) containing mercury, water, and oil. In the state of equilibrium the mercury
is at the bottom, the oil at the top, and the water in the middle; if the

in different planes, and do not intersect each other, there is no metacentre. This indeed
is the case for most of the displacements to which a floating body of irregular shape can be
subjected. There are in general only two directions of heeling to which metacentres correspond, and these two directions are at right angles to each other. For an investigation
of the conditions of stability in floating bodies, see Thomson and Tait's *Natural Philosophy*,
§§ 763–768.

flask is shaken, the liquids are for the moment mixed, but in returning to repose do not fail to resume their former positions.

Fig. 74.
Phial of the Four Elements

It is easily seen from the ordinary rules of hydrostatics, that the surface of separation of two different liquids must be horizontal. Let there be two liquids in a vessel (Fig. 75); the free surface is necessarily horizontal. If now we take two equal superficial elements n and n' in a horizontal layer of the lower liquid, they must be subjected to equal pressures; these pressures are measured by the weights of the liquid cylinders nrs $n'tl$; and these latter cannot be equal unless there be the same height of the lower liquid above the elements n and n'. This reasoning holds for all points in the horizontal layer, which must therefore be at a constant distance from the surface of separation; in other words, this surface must be horizontal.

This property is liable to considerable modification in the case of

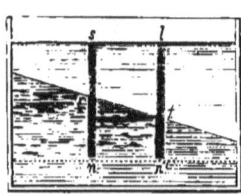

Fig. 75.

liquids which can dissolve each other or act chemically upon each other. Thus, if alcohol be carefully poured upon water in a glass, the two liquids will be seen to have for their surface of junction a horizontal plane; but on agitation a single liquid will be formed by their mutual action, and the separation will not again take place.

If the agitation is not sufficiently great, this intimate mixture will only partially ensue, and will be confined to the neighbourhood of the surface of contact. Two uniform layers of liquid will thus be formed, separated by an intermediate zone of unequal density. This is the case at the mouth of a river, where the fresh water forms on the surface of the sea a layer, the base of which is a compound of fresh and salt water.

CHAPTER X.

81. Determination of Densities.—We have seen in Chap. vii. that in order to determine the density of a body it is only necessary to measure the ratio existing between the weight of a certain volume of the body and the weight of an equal volume of water. The principle of Archimedes enables us to effect this measurement very easily, and the process which it suggests is sometimes more convenient than that which has been described in the chapter mentioned above.

(1.) *Solid bodies.*—Suppose that the object whose density we wish to determine is a piece of copper. It is suspended by a very fine thread to one of the scales of a balance (Fig. 76), its weight is determined, and found to be, say 125·35$^{gr.}$ The body is then immersed in water; the equilibrium is destroyed on account of the upward pressure of the water, and in order to re-establish it, we must add a weight of 14·24$^{gr.}$ to the scale supporting the body. This additional weight, according to the principle of Archimedes, represents the weight of a volume of water equal to the volume of the body. The density of copper is thus $\frac{125\cdot35}{14\cdot24} = 8\cdot8$.

(2.) *Liquid bodies.*—From one of the scales of the balance is suspended (Fig. 77) any body whatever, which must, however, not be capable of being attacked by the liquids in which it is to be immersed; a ball of glass weighted inside with mercury is very well adapted to this purpose. The exact weight of this is obtained; it is then immersed in the liquid whose density is sought—alcohol, for example; an upward pressure is thus produced, and in order to re-establish equilibrium, a weight of 35·43$^{gr.}$ must be added to the scale.

8

The experiment is repeated by immersing the ball in water, in which case the upward pressure is stronger, and a weight of 44·28ᵍʳ· is

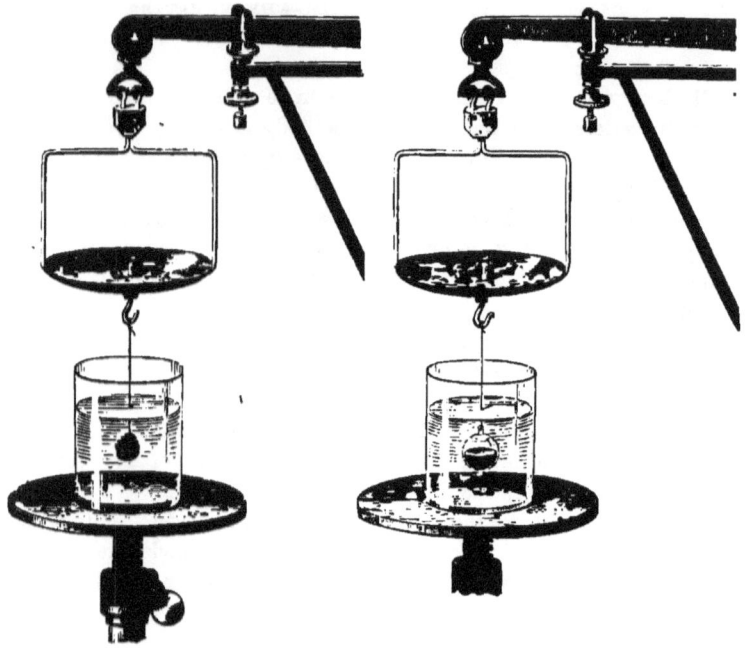

Fig. 76.—Specific Gravity of
Solids.

Fig. 77.—Specific Gravity of
Liquids.

necessary to re-establish equilibrium. The weights 44·28ᵍʳ· and 35·43ᵍʳ· are the weights of equal volumes of water and alcohol; the density of the latter liquid is therefore $\frac{35\cdot43}{44\cdot28}=0\cdot8$.

82. Hydrometers.—The name hydrometer is given to a class of instruments used for determining the densities of liquids by observing either the depths to which they sink in the liquids or the weights required to be attached to them to make them sink to a given depth. According as they are to be used in the latter or the former of these two ways, they are called hydrometers of constant or of variable immersion. The name areometer (from ἀραιός, rare) is used as synonymous with hydrometer, being probably borrowed from the French name of these instruments, *aréomètre*. The hydrometers of constant immersion most generally known are those of Nicholson and Fahrenheit.

83. Nicholson's Hydrometer.—This instrument, which is represented in Fig. 78, consists of a hollow cylinder of metal with conical ends, terminated above by a very thin rod bearing a small dish, and

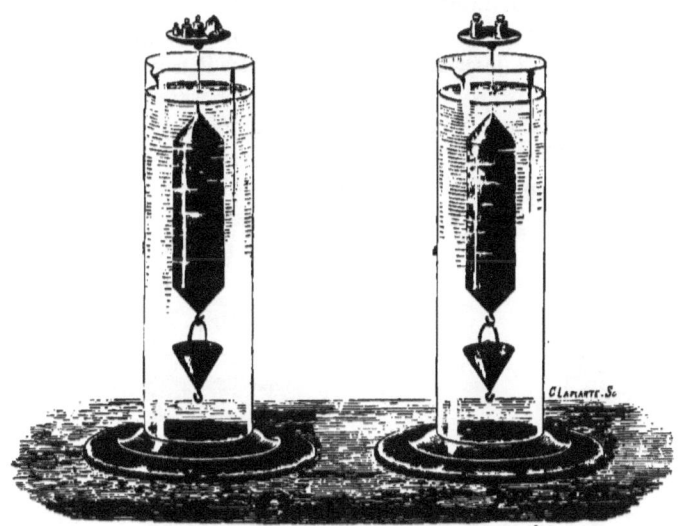

Fig. 78.—Nicholson's Hydrometer.

carrying at its lower end a kind of basket. This latter is of such weight that when the instrument is immersed in water a weight of 1000 grains must be placed in the dish above in order to sink the apparatus as far as a certain mark on the rod. By the principle of Archimedes the weight of the instrument, together with the 1000 grains which it carries, is equal to the weight of the water displaced. Now, let the instrument be placed in another liquid, and the weights in the dish above be altered until they are just sufficient to make the instrument sink to the mark on the rod. If the weights in the dish be called w, and the weight of the instrument itself W, the weight of liquid displaced is now $W+w$, whereas the weight of the same volume of water was $W+1000$; hence the specific gravity of the liquid is $\dfrac{W+w}{W+1000}$.

This instrument can also be used either for weighing small solid bodies or for finding their specific gravities. To find the weight of a body (which we shall suppose to weigh less than 1000 grains), it must be placed in the dish at the top, together with weights just sufficient to make the instrument sink in water as far as the mark. Obviously these weights are the difference between the weight of the body and 1000 grains.

To find the specific gravity of a solid, we first ascertain its weight by the method just described; we then transfer it from the dish above to the basket below, so that it shall be under water during the observation, and observe what additional weights must now be placed in the dish. These additional weights represent the weight of the water displaced by the solid; and the weight of the solid itself divided by this weight is the specific gravity required.

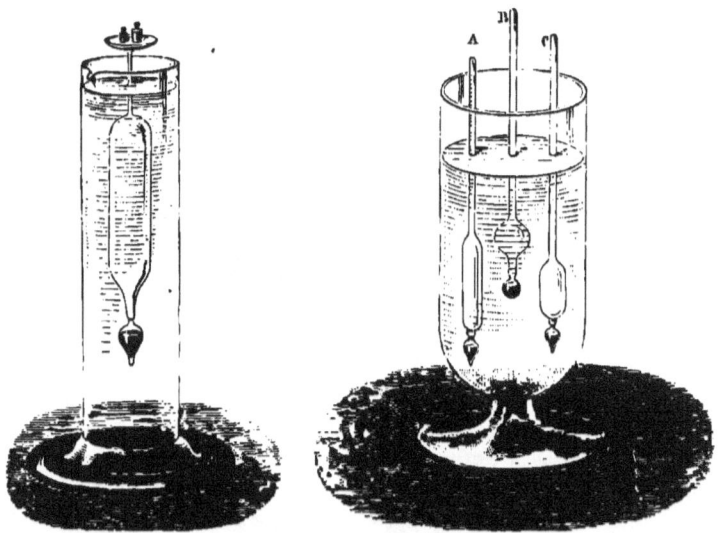

Fig. 79.—Fahrenheit's Hydrometer.   Fig. 80.—Forms of Hydrometers.

84. Fahrenheit's Hydrometer.—This instrument, which is represented in Fig. 79, is generally constructed of glass, and differs from Nicholson's in having at its lower extremity a ball weighted with mercury instead of the basket. It resembles it in having a dish at the top, in which weights are to be placed sufficient to sink the instrument to a definite mark on the stem.

85. Hydrometers of Variable Immersion.—These instruments are usually of the forms represented at A, B, C, Fig. 80. The lower end is weighted with mercury in order to make the instrument sink to a convenient depth and preserve an upright position. The stem is cylindrical, and is graduated, the divisions being frequently marked upon a piece of paper inclosed within the stem, which must in this case be of glass. It is evident that the instrument will sink the deeper the less is the specific gravity of the liquid, since the weight of the liquid displaced must be equal to that of the instrument. Hence if any uniform system of graduation be adopted, so that all the instruments give the same readings in liquids of the same densities, the density of a liquid can be obtained by a mere immersion of the hydrometer—an operation not indeed very precise, but very easy of execution. These instruments have thus come into general use for commercial purposes.

86. General Theory of Hydrometers of Variable Immersion.—Let V be the volume of a hydrometer which is immersed when the instrument floats freely in a liquid whose density (that is, weight per unit volume) is d, then Vd represents the weight of liquid displaced, which by the principle of Archimedes is the same as the weight of the hydrometer itself. If V', d' be the corresponding values for another liquid, we have therefore

$$V d = V' d', \text{ or } d : d' :: V' : V,$$

that is, the density varies inversely as the volume immersed. Let d_1, d_2, d_3...be a series of densities in diminishing order, and V_1, V_2, V_3...the corresponding volumes immersed, which will be in ascending order; then we have

$$d_1, \ d_2, \ d_3 \ldots \text{ proportional to } \frac{1}{V_1}, \ \frac{1}{V_2}, \ \frac{1}{V_3}\ldots$$

$$\text{and } V_1, \ V_2, \ V_3 \ldots \text{ proportional to } \frac{1}{d_1}, \frac{1}{d_2}, \frac{1}{d_3}\ldots$$

Hence, if we wish the divisions to indicate equal differences of density, we must place them so that the corresponding volumes immersed form a harmonical progression. This implies that the divisions must approach nearer together for increasing densities. This is of course on the assumption that the stem is of equal sectional area in all parts as far as the divisions extend.

The following investigation shows how the density of a liquid may be computed from observations made with a hydrometer graduated with equal divisions. It is necessary first to know the divisions to which the instrument sinks in two liquids of known density. Let these divisions be numbered n_1, n_2, reckoning from the top downwards, and let the corresponding densities be d_1, d_2. Now if we take for our unit of volume one of the equal parts on the stem, and if we take c to denote the volume which is immersed when the instrument sinks to the division marked zero, it is obvious that when the instrument sinks to the nth division (reckoned downwards on the stem from zero) the volume immersed is $c-n$, and if the corresponding density be called d, then $(c-n)\,d$ is the weight of the hydrometer. We have therefore

$$(c-n_1)\,d_1=(c-n_2)\,d_2, \text{ whence } c=\frac{n_1 d_1 - n_2 d_2}{d_1 - d_2}.$$

This value of c can be computed once for all.

Then the density D corresponding to any other division N can be found from the equation

$$(c-\text{N})\,\text{D}=(c-n_1)\,d_1 \text{ which gives } \text{D}=\frac{c-n_1}{c-\text{N}},$$

if d_1, which we may suppose to be the density of water, be called unity.

87. Beaumé's Hydrometers.—In these instruments the divisions

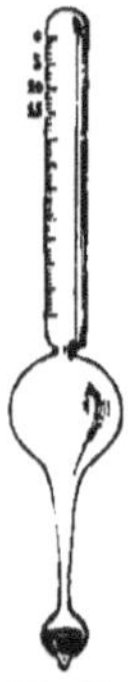

Fig. 81.
Beaumé's Sali-
meter.

are equidistant. There are two distinct modes of graduation, according as the instrument is to be used for determining densities greater or less than that of water. In the former case the instrument is called a salimeter, and is so constructed that when immersed in pure water of the temperature 12° Cent. it sinks nearly to the top of the stem, and the point thus determined is the zero of the scale. It is then immersed in a solution of 15 parts of salt to 85 of water, the density of which is about 1·116, and the point to which it sinks is marked 15. The interval is divided into 15 equal parts, and the graduation is continued to the bottom of the stem, the length of which varies according to circumstances; it generally terminates at the degree 66, which corresponds to sulphuric acid, whose density is commonly the greatest that it is required to determine. Referring to the formulæ of last section, we have here

$$n_1=o, \; d_1=1, \; n_2=15, \; d_2=1·116;$$

whence

$$c=\frac{15\times 1·116}{·116}=144, \; \text{D}=\frac{144}{144-\text{N}}.$$

When the instrument is intended for liquids lighter than water, it is called an alcoholimeter. In this case the point to which it sinks in water is near the bottom of the stem, and is marked 10; the zero of the scale is the point to which it sinks in a solution of 10 parts of salt to 90 of water, the density of which is about 1·085, the divisions in this case being numbered upward from zero.

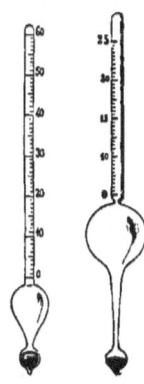

In order to adapt the formulæ of last section to the case of graduations numbered upwards, it is merely necessary to reverse the signs of n_1, n_2, and N; that is we must put

$$c = \frac{n_2 d_2 - n_1 d_1}{d_1 - d_2}, \quad D = \frac{c + n_1}{c + N};$$

and as we have now $n_1 = 10$, $d_1 = 1$, $n_2 = 0$, $d_2 = 1·085$ the formulæ give[1]

$$c = \frac{10}{·085} = 118, \quad D = \frac{128}{118 + N}.$$

Fig. 82. Fig. 83.
Beaumé's Alcoholi-
meters.

87A. Twaddell's Hydrometer.—In this instrument the divisions are placed not as in Beaumé's, at equal distances, but at distances corresponding to equal differences of density. In fact the specific gravity of a liquid is found by multiplying the reading by 5, cutting off three decimal places, and prefixing unity. Thus the degree 1 indicates specific gravity 1·005, 2 indicates 1·010, &c.

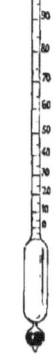

88. Gay Lussac's Centesimal Alcoholimeter.—When a hydrometer is to be used for a special purpose it may be convenient to adopt a mode of graduation different in principle from any that we have described above, and adapted to give a direct indication of the proportion in which two ingredients are mixed in the fluid to be examined. It may indicate, for example, the quantity of salt in sea-water, or the quantity of alcohol in a spirit consisting of alcohol and water. Where there are three or more ingredients of different specific gravities the method fails. Gay-Lussac's alcoholimeter is graduated to indicate, at the temperature of 15° Cent.,

Fig. 87.
Centesimal
Alcoholi-
meter.

---

[1] On comparing the two formulæ for D in this section with the tables in the Appendix to Miller's *Chemical Physics*, I find that as regards the salimeter they agree to two places of decimals and very nearly to three. As regards the alcoholimeter, the table in *Miller* implies that c is about 136, which would make the density corresponding to the zero of the scale about 1·074.

the percentage of pure alcohol in a specimen of spirit. At the top of the stem is 100, the point to which the instrument sinks in pure alcohol, and at the bottom is 0, to which it sinks in water. The position of the intermediate degrees must be determined empirically, by placing the instrument in mixtures of alcohol and water in known proportions, at the temperature of 15°. The law of density, as depending on the proportion of alcohol present, is complicated by the fact that, when alcohol is mixed with water, a diminution of volume (accompanied by rise of temperature) takes place.   .

88 A. Specific Gravity of Mixtures.—When two or more substances are mixed without either shrinkage or expansion (that is, when the volume of the mixture is equal to the sum of the volumes of the components), the density of the mixture can easily be expressed in terms of the quantities and densities of the components.

First, let the volumes v_1, v_2, v_3 . . . of the components be given, together with their densities d_1, d_2, d_3 . . .
Then their masses (or weights) are $v_1 d_1$, $v_2 d_2$, $v_3 d_3$ . . .
The mass of the mixture is the sum of these masses, and its volume is the sum of the volumes v_1, v_2, v_3 . . . ; hence its density is

$$\frac{v_1 d_1 + v_2 d_2 + \ldots}{v_1 + v_2 + \ldots}.$$

Secondly, let the weights or masses m_1, m_2, m_3 . . . of the components be given, together with their densities d_1, d_2, d_3 . . .
Then their volumes are $\frac{m_1}{d_1}$, $\frac{m_2}{d_2}$, $\frac{m_3}{d_3}$ . . .
The volume of the mixture is the sum of these volumes, and its mass is $m_1 + m_2 + m_3 + $ . . . ; hence its density is

$$\frac{m_1 + m_2 + \ldots}{\frac{m_1}{d_1} + \frac{m_2}{d_2} + \ldots}.$$

88 B. Graphical Method of Graduation.—When the points on the stem which correspond to some five or six known densities, nearly equidifferent, have been determined, the intermediate graduations can be inserted with tolerable accuracy by the graphical method of interpolation, a method which has many applications in physics besides that which we are now considering. Suppose A and B (Fig. 85) to represent the extreme points, and I, K, L, R intermediate points, all of which correspond to known densities. Erect ordinates (that is to say, perpendiculars) at these points, proportional to the respective densities, or (which will serve our purpose equally well)

erect ordinates II', KK', LL', RR', BC proportional to the excesses of
the densities at I, K, L, R, B above the density at A.   Any scale of

cqual parts can be employed
for laying off the ordinates,
but it is convenient to adopt
a scale which will make the
greatest ordinate BC not
much greater nor much less
than the base line AB. In
the figure, the density at B is
supposed to be 1·80, the density at A being 1.   The difference of density is therefore
·80, as indicated by the figures 80 on the scale of equal
parts.

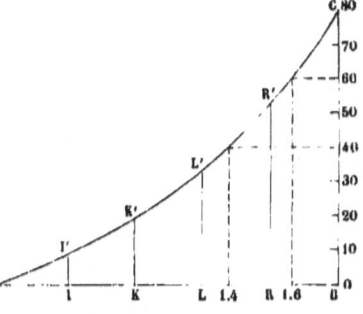

Fig. 85.—Graphical Method of Graduation.

Having erected the ordinates, we must draw through their
extremities the curve AI'K'L'R'C, making it as free from sudden
turns as possible, as it is upon the regularity of this curve that the
accuracy of the interpolation depends. Then to find the point on the
stem AB at which any other density is to be marked—say 1·60, we
must draw through the 60th division, on the line of equal parts, a
horizontal line to meet the curve, and, through the point thus found
on the curve, draw an ordinate. This ordinate will meet the base
line AB in the required point, which is accordingly marked 1·6 in
the figure. The curve also affords the means of solving the converse
problem, that is, of finding the density corresponding to any given
point on the stem. At the given point in AB, which represents the
stem, we must draw an ordinate, and through the point where this
meets the curve we must draw a horizontal line to meet the scale of
equal parts. The point thus determined on the scale of equal parts
indicates the density required, or rather the excess of this density
above the density of A.

CHAPTER XI.

89. Equilibrium in Vessels in Communication.—When a liquid is contained in vessels communicating with each other, and is in equilibrium, it stands at the same height in the different parts of the system, so that the free surfaces all lie in the same horizontal plane.

This is an immediate consequence of the fact that layers of equal

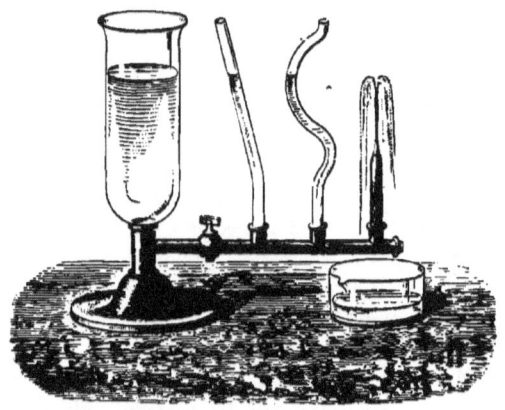

Fig. 89.—Vessels in Communication.

pressure in a liquid are always horizontal (§ 64); for if we take any such layer at the bottom of the system, we must proceed upwards through the same vertical height in all parts of the system in order to reach the free surface which corresponds to the pressure. Thus, in the system represented by Fig. 89, the liquid is seen to stand at the same height in the principal vessel and in the variously shaped tubes communicating with it. If one of these tubes is cut off at a height less than that of the liquid in the principal vessel, and if it be made to terminate in a narrow mouth, the liquid will be seen to spout up nearly to the level of that in the principal vessel.

This tendency of liquids to find their own level is very important, and of continual application. Thus, a reservoir of water may have different pipes issuing from it and spreading out in all possible directions with any number of turns and windings; provided that the ends of these pipes lie below the level of the reservoir, the water will flow through the pipes and run out at their extremities. The velocity of exit, however, will depend on the form and arrangement of the pipes, as well as on the difference of level. This velocity must of course be taken into account in calculating the quantity of water that will flow in a given time; and in forming plans for the proper distribution of public supplies of water. It also determines the height to which a jet of water can be discharged from an opening at the end of the pipe.

90. Water-level.—The well-known instrument called the water-level depends upon the property just mentioned. It consists of a

Fig. 90.—Water-level.

metal tube *bb*, bent at right angles at its extremities. These carry two glass tubes *aa*, very narrow at the top, and of the same diameter. The tube rests on a tripod stand, at the top of which is a joint that enables the observer to turn the apparatus and set it in any direction. The tube is placed in a position *nearly* horizontal, and water, generally coloured a little, is poured in until it stands at about three-fourths of the height of each of the glass tubes.

By the principle of equilibrium in vessels communicating with each other, the surfaces of the liquid in the two branches are in the same horizontal plane, so that if the line of the observer's sight just grazes the two surfaces, it will be horizontal.

This is the principle of the operation called *levelling*, the object of which is to determine the difference of vertical height, or *difference of level*, between two given points.   Suppose A and B to be the two points (Fig. 91).   At each of these points is fixed a levelling-staff,

Fig. 91.—Levelling.

that is, an upright rod divided into parts of equal length, on which slides a small square board whose centre serves as a mark for the observer.

The level being placed at an intermediate station, the observer directs the line of sight towards each levelling-staff, and the mark is raised or lowered till the line of sight passes through its centre. The marks on the two staves are in this way brought to the same level.   The staff in the rear is then carried in advance of the other, the level is again placed between the two, and another observation taken.   In this way, by noting the division of the staff at which the sliding mark stands in each case, the difference of levels of two distant stations can be deduced from observations at a number of intermediate points.

91. Spirit-level.—These observations can be made in a much more exact and convenient manner by means of the spirit-level.   This

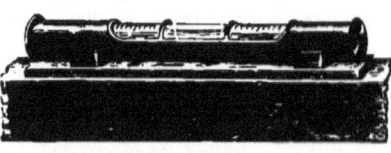

Fig. 92.—Spirit-level.

instrument is composed of a glass tube slightly curved, containing a liquid, which is generally alcohol, and which fills the whole extent of the tube, except a small space occupied by an air-bubble. This tube is inclosed in a mounting which is firmly supported on a stand.

Suppose the tube to have been so constructed that a vertical section of its upper surface is an arc of a circle, and suppose the instrument placed upon a horizontal plane (Fig. 93). The air-bubble will take up a position MN at the highest part of the tube, such that the arcs MA and NB are equal. Hence it follows that if the level be reversed end for end, the bubble will occupy the same position, the point N coming to M, and *vice versa*. This will not be the case if AB is inclined to the horizon (Fig. 94), for then the distance MA being different from NB, after the apparatus has been turned, the bubble will assume a symmetrical position at the opposite end of the tube. The condition,

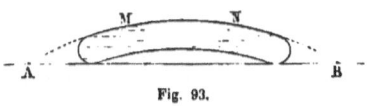

Fig. 93.

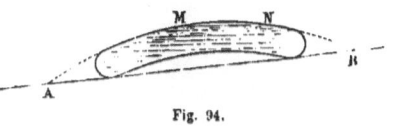

Fig. 94.

therefore, that the line on which the spirit-level rests should be horizontal is, that after this operation of reversal the bubble should remain within the same limits. In order to avoid the trouble of turning the instrument, the maker marks these limits by reference-marks on the tube or its mounting, and in order to determine that a line is horizontal it is only necessary to make sure that, when the level is placed upon it, the bubble lies exactly between these reference-marks.

In order that a plane surface may be horizontal, we must have two lines in it horizontal. This result is practically attained in the

Fig. 95.—Testing the Horizontality of a Surface.

following manner:—The surface is made to rest on three levelling screws which form the three vertices of an isosceles triangle; the level is first placed parallel to the base of the triangle, and, by means of one of the screws, the bubble is brought between the reference-marks. The instrument is then placed perpendicularly to its first position, and the bubble is brought between the marks by means of the third screw; this second operation cannot disturb the result of the first, since the plane has only been turned about a horizontal line as hinge.

92. Level furnished with Telescope.—In order to apply the spirit-level to land-surveying, an apparatus such as that represented

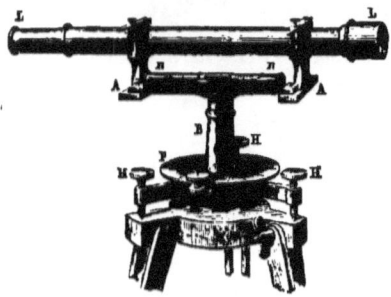

in the figure is employed. Upon a frame **AA**, movable about a vertical axis B, are placed a spirit-level *nn*, and a telescope LL, in positions parallel to each other. The telescope is furnished at its focus with two fine wires crossing one another, whose point of intersection determines the line of sight with great precision. The appara-

Fig. 96.—Spirit-level with Telescope.

tus, which is provided with levelling screws H, rests on a tripod stand, and the observer is able, by turning it about its axis, to command the different points of the horizon. By a process of adjustment which need not here be described, it is known that when the bubble is between the marks the line of sight is horizontal; so that we may proceed to find the difference of level between two points in the same way as with the water-level; but the operation is much more precise, and the range of vision much more extensive. By furnishing the instrument with a graduated horizontal circle P, we may obtain the azimuths of the points observed, and thus map out contour lines.

On each side of the reference-marks of the bubble are divisions for measuring small deviations from horizontality. It is, in fact, easy to see, by reference to Fig. 93, that by tilting the level through any small angle, the bubble is displaced by a quantity proportional to this angle, at least when the curvature of the instrument is that of a circle.

For determining the angular value corresponding to each division of the tube, it is usual to employ an apparatus opening like a pair of compasses by a hinge C, on one of the legs of which rests, by two V-shaped supports, the tube T of the level. The compass is opened by means of a micrometer screw V, of very regular action; and as the distance of the screw from the hinge is known, as well as the distance between the threads of the screw, it is easy to calculate beforehand the value of the divisions on the micrometer head. The levelling screws of the instrument serve to bring the bubble between

Fig. 97.—Graduation of Spirit-level.

its reference-marks, so that the micrometer screw is only used to determine the value of the divisions on the tube.

93. Equilibrium of Two Different Liquids in Communicating Vessels. —If into one of two tubes in communication we pour a liquid, say mercury, this liquid will rise to the same height in both branches. If we now pour water into one of them, the mercury will be pushed back in the other branch; and when equilibrium has been established, the *heights of the two liquids above the surface of separation* will be very unequal, as shown in the figure. In general, these heights, since they correspond to the same pressure upon the surface of separation, *will be inversely proportional to the densities.*

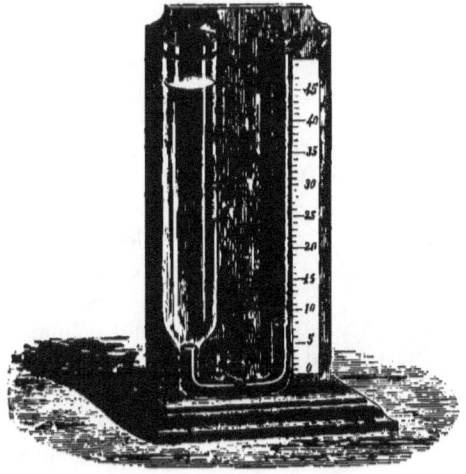

Fig. 98.—Equilibrium of Two Fluids in Communicating Vessels.

94. Capillarity—General Phenomena.—The different principles of equilibrium which have been explained in the preceding paragraphs, are subject to remarkable exceptions when the vessels in

which the liquids are contained are very narrow, or, as they are
called, capillary (*capillus*, a hair); and also in the case of vessels of
any size, when we consider the portion of the liquid which is in
close proximity to the sides.

1. *Free Surface.*—The surface of a liquid is not horizontal in the
neighbourhood of the sides of the vessel, but presents a very decided
curvature.   When the liquid wets the vessel, as in the case of water
in a glass vessel (Fig. 99), the surface is concave; on the contrary
when the liquid does not wet the vessel, as in the case of mercury in
a glass vessel (Fig. 100), the surface is, generally speaking, convex.

2. *Capillary Elevation and Depression.*—If a very narrow tube
of glass be plunged in water, or any other liquid that will wet it

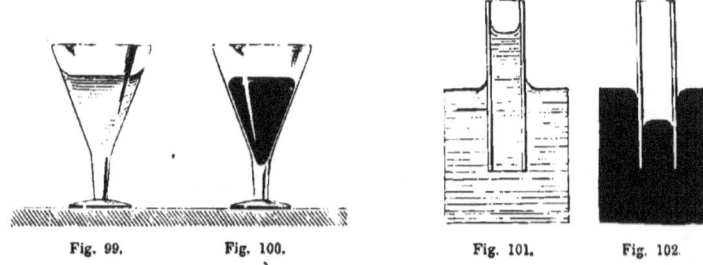

Fig. 99.   Fig. 100.   Fig. 101.   Fig. 102.

(Fig. 101), it will be observed that the level of the liquid, instead of
remaining at the same height inside and outside of the tube, stands
perceptibly higher in the tube; a *capillary ascension* takes place,

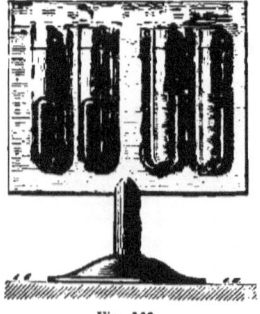

Fig. 103.

which varies in amount according to the
nature of the liquid and the diameter of
the tube.   It will also be seen that the
liquid column thus raised terminates in
a concave surface.   If a glass tube be
dipped in 'mercury, which does not wet
it, it will be seen, by bringing the tube
to the side of the vessel, that the mercury
is depressed in its interior, and that it
terminates in a convex surface (Fig. 102).

3. *Capillary Vessels in Communication with Others.*—If we take two bent
tubes, each having one branch of a considerable diameter and the
other extremely narrow, and pour into one of them a liquid which

wets it, and into the other mercury, the liquid will be observed in the former case to stand higher in the capillary than in the principal branch, and in the latter case to stand lower; the free surfaces being at the same time concave in the case of the liquid which wets the tubes, and convex in the case of the mercury.

95. Circumstances which influence Capillary Elevation and Depression. —In wetted tubes the elevation depends upon the nature of the liquid; thus, at the temperature of 18° Cent., water rises 29.79^{mm} (1.2 inch) in a tube 1 millimetre ($\frac{1}{25}$ inch) in diameter, alcohol rises 12.18^{mm}, nitric acid 22.57^{mm}, essence of lavender 4.28^{mm}, &c. The nature of the tube is almost entirely immaterial, provided the precaution be first taken of wetting it with the liquid to be employed in the experiment, so as to leave a film of the liquid adhering to the sides of the tube.

Capillary depression, on the other hand, depends both on the nature of the liquid and on that of the tube. Both ascension and depression diminish as the temperature increases; for example, the elevation of water, which in a tube of a certain diameter is equal to 132^{mm} at 0° Cent., is only 106^{mm} at 100°.

96. Law of Diameters.—*Capillary elevations and depressions, when all other circumstances are the same, are inversely proportional to the diameters of the tubes.* As this law is a consequence of the mathematical theories which are generally accepted as explaining capillary phenomena, its verification has been regarded as of great importance.

The experiments of Gay-Lussac, which confirmed this law, have been repeated, with slight modifications, by several observers. The method employed consists essentially in measuring the capillary elevation of a liquid by means of a cathetometer (Fig. 104). The telescope ll is directed first to the top n of the column in the tube, and then to the end of a pointer b, which touches the surface of the liquid at a point where it is horizontal. In observing the depression of mercury, since the opacity of the metal prevents us from seeing the tube, we must bring the tube close to the side of the vessel e.

The diameter of the tube can be measured directly by observing its section through a microscope, or we may proceed by the method employed by Gay-Lussac. He weighed the quantity of mercury which filled a known length l of the tube; this weight w is that of a cylinder of mercury whose radius x is determined by the equation $13.59\ \pi x^2 l = w.$*

* This formula is only true for the metrical system, 13.59 being the specific gravity of mercury. If x and l are in centimetres, w will be in grammes.

The result of these different experiments is, that in the case of
wetted tubes the law is exactly fulfilled, provided that they be previously washed with the greatest care, so as to remove all foreign
matters, and that the liquid on which the experiment is to be per-

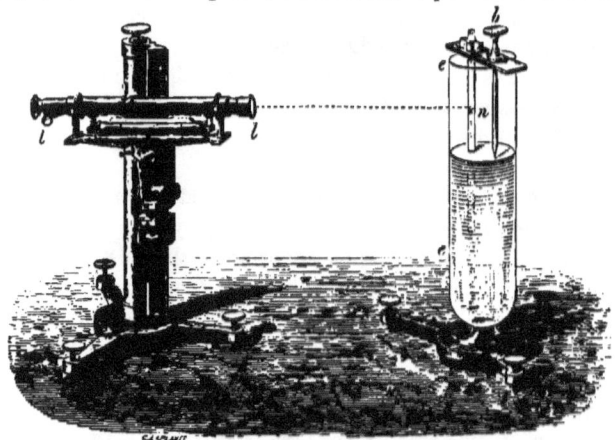

Fig. 104.—Verification of Law of Diameters.

formed be first passed through them. When the liquid does not wet
the tube, various causes combine to affect the form of the surface in
which the liquid column terminates; and we cannot infer the depression from knowing the diameter, unless we also take into consideration some element connected with the form of the terminal surface,
such as the length of the sagitta, or the angle made with the sides
of the tube by the extremities of the curved surface, which is called
the *angle of contact*.

97. Cause of Capillary Phenomena.—Capillary phenomena, as they
take place alike in air and in vacuo, cannot be attributed to the
action of the atmosphere. They depend upon molecular actions
which take place between the particles of the liquid itself, and between the liquid and the solid containing it, the actions in question
being purely superficial—that is to say, being confined to an extremely
thin layer forming the external boundary of the liquid, and to an
extremely thin superficial layer of the solid in contact with the
liquid. For example, it is found in the case of glass tubes, that the
amount of capillary elevation or depression is not at all affected by
the thickness of the sides of the tube. The following are some of
the principles which govern capillary phenomena.

1. For a given liquid in contact with a given solid, with a definite intimateness of contact (this last element being dependent upon the cleanness of the surface, upon whether the surface of the solid has been recently washed by the liquid, and perhaps upon some other particulars), there is (at any specified temperature) a definite angle of contact, which is independent of the directions of the surfaces with regard to the vertical.

2. Every liquid behaves as if a thin film, forming its external layer, were in a state of tension, and exerting a constant effort to contract. This tension, or contractile force, is exhibited over the whole of the free surface (that is, the surface which is exposed to air), but wherever the liquid is in contact with a solid, its existence is masked by other molecular actions. It is uniform in all directions in the free surface, and at all points in this surface, being dependent only on the nature and temperature of the liquid. Its intensity for several specified liquids is given in tabular form further on (§ 97F) upon the authority of Van der Mensbrugghe. Tension of this kind must of course be stated in units of force per linear unit, because by doubling the width of a band we double the force required to keep it stretched. Mensbrugghe considers that such tension really exists in the superficial layer; but the majority of authors (and we think with more justice) regard it rather as a convenient fiction, which accurately represents the effects of the real cause. Two of the most eminent writers on the cause of capillary phenomena are Laplace and Dr. Thomas Young. The subject presents difficulties which have not yet been fully surmounted.

The law of diameters is a direct consequence of the two preceding principles; for if a denote the external angle of contact (which is acute in the case of mercury against glass), T the tension per unit length, and r the radius of the tube, then $2\pi r$T will be the whole amount of force exerted at the margin of the surface; and as this force is exerted in a direction making an angle a with the vertical, its vertical component will be $2\pi r$T$\cos a$, which is exerted in pulling the tube upwards and the liquid downwards.

If w be the weight of unit volume of the liquid, then $\pi r^2 w$ is the weight of as much as would occupy unit length of the tube; and if h denote the height of a column whose weight is equal to the force tending to depress the liquid, we have

$$\pi r^2 h w = 2\pi r \text{T} \cos a;$$

whence $h = \dfrac{2\,T\cos a}{r\,.\,w}$, which, when the other elements are given, varies inversely as r, the radius of the tube.

Having regard to the fact that the surface is not of the same height in the centre as at the edges, it is obvious that h denotes the mean height.

If a be obtuse, h will be negative—that is to say, there will be elevation instead of depression. In the case of water against a tube which has been well wetted with that liquid, a is 180°—that is to say, the tube is tangential to the surface. For this case the formula for h gives

$$\text{elevation} = \frac{2\,T}{r\,w}.$$

Again, for two parallel vertical plates at distance u, the vertical force of capillarity for a unit of length is $2\,T\cos a$, which must be equal to $w\,h\,u$, being the weight of a sheet of liquid of height h, thickness u, and length unity. We have therefore

$$h = \frac{2\,T\cos a}{u\,w},$$

which agrees with the expression for the depression or elevation in a circular tube whose radius is equal to the distance between these parallel plates.

The surface tension always tends to reduce the surface to the smallest area which can be inclosed by its actual boundary; and therefore always produces a normal force directed *towards* the concave side of the superficial film. Hence, wherever there is capillary elevation the free surface must be concave; wherever there is depression it must be convex.

97A. It follows from a well-known proposition in statics (Todhunter's *Statics*, § 194), that if a *cylindrical* film be stretched with a uniform tension T (so that the force tending to pull the film asunder across any short line drawn on the film, is T times the length of the line), the resultant normal pressure (which the film exerts, for example, against the surface of a solid internal cylinder over which it is stretched) is T divided by the radius of the cylinder.

It can be proved that a film of any form, stretched with uniform tension T, exerts at each point a normal pressure equal to the sum of the pressures which would be exerted by two overlapping cylindrical films, whose axes are at right angles to one another, and whose cross sections are circles of curvature of normal sections at the point. That is to say, if P be the normal force per unit area, and $r\,r'$ the radii of curvature in two mutually perpendicular normal sections at the point, then

$$P = T\left(\frac{1}{r} + \frac{1}{r'}\right).$$

At any point on a curved surface, the normal sections of greatest and least curvature are mutually perpendicular, and are called the principal normal sections at the point. If the corresponding radii of curvature be R, R', we have

$$P = T\left(\frac{1}{R} + \frac{1}{R'}\right); \qquad (1)$$

or *the normal force per unit area is equal to the tension per unit length multiplied by the sum of the principal curvatures.*

In the case of capillary depressions and elevations, the superficial film at the free surface is to be regarded as pressing the liquid inwards, or pulling it outwards, according as this surface is convex or concave, with a force P given by the above formula. The value of P at any point of the free surface is equal to the pressure due to the height of a column of liquid extending from that point to the level of the general horizontal surface. It is therefore greatest at the edges of the elevated or depressed column in a tube, and least in the centre; and the curvature, as measured by $\frac{1}{R} + \frac{1}{R'}$, must vary in the same proportion. If the tube is so large that there is no sensible elevation or depression in the centre of the column, the centre of the free surface must be sensibly plane.

97B. Another consequence of the formula is, that in circumstances where there can be no normal pressure towards either side of the surface,

$$\frac{1}{R} + \frac{1}{R'} = 0; \qquad (2)$$

which implies that either the surface is plane, in which case each of the two terms is separately equal to zero, or else

$$R = -R'; \qquad (3)$$

that is, the principal radii of curvature are equal, and lie on opposite sides of the surface. The formulæ (2), (3) apply to a film of soapy water attached to a loop of wire. If the loop be in one plane, the film will be in the same plane. If the loop be not in one plane, the film cannot be in one plane, and will in fact assume that form which gives the least area consistent with having the loop for its boundary. At every point it will be observed to be, if we may so say, concave towards both sides, and convex towards both sides, the concavity being precisely equal to the convexity—that is to say, equation (3) is satisfied at every point of the film.

In this case both sides of the film are exposed to atmospheric pressure. In the case of a common soap-bubble the outside is exposed to atmospheric pressure, and the inside to a pressure somewhat greater, the difference of the pressures being compensated by the tendency of the film to contract. Formula (1) becomes for either the outer or inner surface of a spherical bubble

$$P = \frac{2T}{R};$$

but this result must be doubled, because there are two free surfaces; hence the excess of pressure of the inclosed above the external air is $\frac{4T}{R}$, R denoting the radius of the bubble.

The value of T for soapy water is about 1 grain per linear inch; hence, if we divide 4 by the radius of the bubble expressed in inches, we shall obtain the excess of internal over external pressure *in grains per square inch.*

The value of T for any liquid may be obtained by observing the amount of elevation or depression in a tube of given diameter, and employing the formula

$$T = \frac{whr}{2\cos a}, \qquad (4)$$

which follows immediately from the formula for h in § 97.

97 c. It is this uniform surface tension, of which we have been speaking, which causes a drop of a liquid falling through the air either to assume the spherical form, or to oscillate about the spherical form. The phenomena of drops can be imitated on an enlarged scale, under circumstances which permit us to observe the actual motions, by a method devised by Professor Plateau of Ghent. Olive-oil is intermediate in density between water and alcohol. Let a mixture of alcohol and water be prepared, having precisely the density of olive-oil, and let about a cubic inch of the latter be gently introduced into it with the aid of a funnel or pipette. It will assume a spherical form, and if forced out of this form and then left free, will slowly oscillate about it; for example, if it has been compelled to assume the form of a prolate spheroid, it will pass to the oblate form, will then become prolate again, and so on alternately, becoming however more nearly spherical every time, because its movements are hindered by friction, until at last it comes to rest as a sphere.

97 D. Capillarity furnishes no exception to the principle that the pressure in a liquid is the same at all points at the same depth. When the free surface within a tube is convex, and is consequently depressed below the general level of the external surface, the pressure becomes suddenly greater on passing downwards through the superficial layer, by the amount due to the curvature. Below this it increases regularly by the amount due to the depth of liquid passed through. The pressure at any point vertically under the convex meniscus[1] may be computed, either by taking the depth of the point below the general free surface, and adding atmospheric pressure to the pressure due to this depth, according to the ordinary principles of hydrostatics, or by taking the depth of the point below that point of the meniscus which is vertically over it, adding the pressure due to the curvature at this point, and also adding atmospheric pressure.

When the free surface of the liquid within a tube is concave, the pressure suddenly diminishes on passing downwards through the superficial layer, by the amount due to the curvature as given by formula (1); that is to say, the pressure at a very small depth is less than atmospheric pressure by this amount. Below this depth it goes on increasing according to the usual law, and becomes equal to atmospheric pressure at that depth which corresponds with the level of the general external surface. The pressure at any point in the liquid within the tube can therefore be obtained either by subtracting from atmospheric pressure the pressure due to the elevation of the point above the general surface, or by adding to atmospheric pressure the pressure due to the depth below that point of the meniscus which is on the same vertical, and subtracting the pressure due to the curvature at this point.

These rules imply, as has been already remarked, that the curvature is different at different points of the meniscus, being greatest where the elevation or depression is greatest, namely at the edges of the meniscus; and least at the point of least elevation or depression, which in a cylindrical tube is the middle point.

---

[1] The convex or concave surface of the liquid in a tube is usually denoted by the name *meniscus* (μηνίσκος, a crescent), which denotes a form approximately resembling that of a watch-glass.

The principles just stated apply to all cases of capillary elevation and depression.

They enable us to calculate the force with which two parallel vertical plates, partially immersed in a liquid which wets them, are urged towards each other by capillary action. The portion of liquid elevated between them is at less than atmospheric pressure, and therefore is insufficient to resist the atmospheric pressure which is exerted on the outer faces of the plates. The average pressure in the elevated portion of liquid is that which exists half-way up it, and is less than atmospheric pressure by the pressure of a column of liquid whose height is half the elevation.

Even if the liquid be one which does not wet the plates, they will still be urged towards each other by capillary action; for the inner faces of the plates are exposed to merely atmospheric pressure over that portion of their areas which corresponds to the depression, while the corresponding portions of the external faces are exposed to atmospheric pressure increased by the weight of a portion of the liquid.

These principles explain the apparent attraction exhibited by bodies floating on a liquid which either wets them both or wets neither of them. When the two bodies are near each other they behave somewhat like parallel plates, the elevation or depression of the liquid between them being greater than on their remote sides.

If two floating bodies, one of which is wetted and the other unwetted by the liquid, come near together, the elevation and depression of the liquid will be less on the near than on the remote sides, and apparent repulsion will be exhibited.

In all cases of capillary elevation or depression, the solid is pulled downwards or upwards with a force equal to that by which the liquid is raised or depressed. In applying the principle of Archimedes to a solid partially immersed in a liquid, it is therefore necessary (as we have seen in § 79), when the solid produces capillary depression, to reckon the void space thus created as part of the displacement; and when the solid produces capillary elevation, the fluid raised above the general level must be reckoned as *negative* displacement, tending to *increase* the apparent weight of the solid.

97E. Thus far all the effects of capillary action which we have mentioned are connected with the curvature of the superficial film, and depend upon the principle that a convex surface increases and a concave surface diminishes the pressure in the interior of the liquid.

But there is good reason for maintaining that whatever be the form of the free surface there is always a certain amount of pressure in the interior due to the molecular action at this surface, and that the pressure due to the curvature of the surface is to be added to or subtracted from a definite amount of pressure which is independent of the curvature and depends only on the nature and condition of the liquid. This indeed follows at once from the fact that capillary elevation can take place in vacuo. As far as the principles of the preceding paragraphs are concerned, we should have, at points within the elevated column, a pressure less than that existing in the vacuum. This, however, cannot be; we cannot conceive of negative pressure existing in the interior of a liquid, and we are driven to conclude that the elevation is owing to the excess of the pressure caused by the plane surface in the containing vessel above the pressure caused by the concave surface in the capillary tube.

There are some other facts which seem only explicable on the same general principle of interior pressure due to surface action,—facts which attracted the notice of some of the earliest writers on pneumatics, namely, that siphons will work in vacuo, and that a column of mercury at least 75 inches in length can be sustained—as if by atmospheric pressure—in a barometer tube, the mercury being boiled and completely filling the tube.

97F. We have now to notice certain phenomena which depend on the difference in the surface tensions of different liquids, or of the same liquid in different states.

Let a thin layer of oil be spread over the upper surface of a thin sheet of brass, and let a lamp be placed underneath. The oil will be observed to run away from the spot directly over the flame, even though this spot be somewhat lower than the rest of the sheet. This effect is attributable to the excess of surface tension in the cold oil above the hot.

In like manner, if a drop of alcohol be introduced into a thin layer of water spread over a nearly horizontal surface, it will be drawn away in all directions by the surrounding water, leaving a nearly dry spot in the space which it occupied. In this experiment the water should be coloured in order to distinguish it from the alcohol.

Again, let a very small fragment of camphor be placed on the surface of hot water. It will be observed to rush to and fro with frequent rotations on its own axis, sometimes in one direction and sometimes in the opposite. These effects, which have been a frequent subject of discussion, are now known to be due to the diminution of the surface tension of the water by the camphor which it takes up. Superficial currents are thus created, radiating from the fragment of camphor in all directions; and as the camphor dissolves more quickly in some parts than in others, the currents which are formed are not equal in all directions, and those which are most powerful prevail over the others and give motion to the fragment.

The values of T, the apparent surface tension, for several liquids, are given in the following table, on the authority of Van der Mensbrugghe, in milligrammes (or thousandth parts of a gramme) per millimetre of length. They can be reduced to grains per inch of length by multiplying them by ·392; for example, the surface tension of distilled water is $7·3 \times ·392 = 2·86$ grains per inch.

Distilled water at 20° Cent.,	7·3	Solution of Marseilles soap, 1 part of	
Sulphuric ether,	1·88	soap to 40 of water,	2·83
Absolute alcohol,	2·5	Solution of saponine,	4·67
Olive-oil,	3·5	Saturated solution of carbonate of	
Mercury,	49·1	soda,	4·28
Bisulphide of carbon,	3·57	Water impregnated with camphor,	4·5

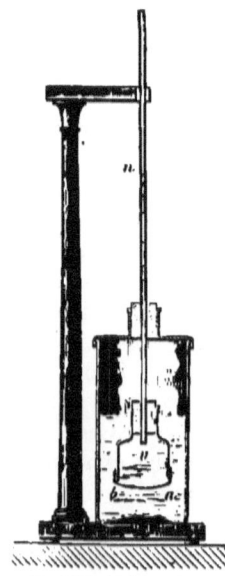

Fig. 105.—Endosmometer.

98. Endosmose.—Capillary phenomena have undoubtedly some connection with a very important property discovered by Dutrochet, and called by him *endosmose*.

The *endosmometer* invented by him to illustrate this phenomenon consists of a reservoir *v* closed below by a membrane *ba*, and terminating above in a tube of considerable length. This reservoir is filled, suppose, with a solution of gum in water, and is kept immersed in water. At the end of some time the level of the liquid in the tube will be observed to have risen to *n*, suppose, and at the same time traces of gum will be found in the water in which the reservoir is immersed. Hence we conclude that the two liquids have penetrated through the membrane, but in different proportions; and this is what is called endosmose.

If instead of a solution of gum we employed water containing albumen, sugar, or gelatine in solution, a similar result would ensue. The membrane may be replaced by a slab of wood or of porous clay. Physiologists have justly attached very great importance to this discovery of Dutrochet. It explains, in fact, the interchange of liquids which is continually taking place in the tissues and vessels of the animal system, as well as the absorption of water by the spongioles of roots, and several similar phenomena.

As regards the power of passing through porous diaphragms, Graham has divided substances into two classes—*crystalloids* and *colloids* (κόλλη, glue). The former are susceptible of crystallization, form solutions free from viscosity, are sapid, and possess great powers of diffusion through porous septa. The latter, including gum, starch, albumen, &c., are characterized by a remarkable sluggishness and indisposition both to diffusion and to crystallization, and when pure are nearly tasteless.

CHAPTER XII.

99. Weight of the Air and of Gases.—Gaseous bodies possess a number of properties in common with liquids; like them, they transmit pressures entire and in all directions, according to the principle of Pascal; but they differ essentially from liquids in the permanent repulsive force exerted between their molecules, in virtue of which a mass of gas always tends to expand.

The opinion was long held that the air was without weight; or, to speak more precisely, it never occurred to any of the philosophers who preceded Galileo to attribute any influence in natural phenomena to the weight of the air. And as this influence is really of the first importance, and comes into play in many of the commonest phenomena, it very naturally happened that the discovery of the weight of air formed the commencement of the modern revival of physical science.

It appears, however, that Aristotle conceived the idea of the possibility of air having weight, and, in order to convince himself on this point, he weighed a skin inflated and collapsed. As he obtained the same weight in both cases, he relinquished the idea which he had for the moment entertained. In fact, the experiment, as he performed it, could only give a negative result; for if the weight of the skin was increased, on the one hand, by the introduction of a fresh quantity of air, it was diminished, on the other, by the corresponding increase in the upward pressure of the air displaced. In order to draw a certain conclusion, the experiment should be performed with a vessel which could receive within it air of different degrees of density, without changing its own volume.

Galileo is said to have devised the experiment of weighing a globe filled alternately with ordinary air and with compressed air.

As the weight is greater in the latter case, Galileo should have drawn the inference that air is heavy. It does not appear, however, that the importance of this conclusion made much impression on him, for he did not give it any of those developments which might have been expected to present themselves to a mind like his.

100. Experiment of Otto Guericke.—Otto Guericke, the illustrious inventor of the air-pump, in 1650 performed the following experiment, which is decisive:—

A globe of glass, furnished with a stop-cock, and of a sufficient capacity (about twelve litres), is exhausted of air. It is then suspended from one of the scales of a balance, and a weight sufficient to produce equilibrium is placed in the other scale. The stop-cock is then opened, the air rushes into the globe, and the beam is observed gradually to incline, so that an additional weight is required in the other scale, in order to re-establish equilibrium. If the capacity of the globe is 12 litres, about 15·5 grammes will be needed, which gives 1·3 gramme as the approximate weight of a litre of air.[1]

Fig. 106.— Weight of Air.

If, in performing this experiment, we take particular precautions to insure its precision, as we shall explain in the book on heat, it will be found that, at the temperature of freezing water, and under the pressure of one atmosphere, a litre of air weighs 1·293 gramme.[2] Under these circumstances, the ratio of the weight of a volume of air to that of an equal volume of water is $\frac{1\cdot293}{1000} = \frac{1}{773}$. Air is thus 773 times lighter than water.

By repeating this experiment with other gases, we may determine

[1] A cubic foot of air in ordinary circumstances weighs about an ounce and a quarter.

[2] In strictness, the weight in grammes of a litre of air under the pressure of 760 millimetres of mercury is different in different localities, being proportional to the intensity of gravity—not because the force of gravity on the litre of air is different, for though this

their weight as compared with that of air, and the absolute weight of a litre of each of them. Thus it is found that a litre of oxygen weighs 1·43 gramme, a litre of carbonic acid 1·97 gramme, a litre of hydrogen 0·089 gramme, &c.

101. Atmospheric Pressure.—The atmosphere encircles the earth with a layer some 50 or 100 miles in thickness; this heavy fluid mass exerts on the surface of all bodies a pressure entirely analogous both in nature and origin to that sustained by a body wholly immersed in a liquid. It is subject to the fundamental law mentioned in § 64. The pressure should therefore diminish as we ascend from the surface of the earth, but should have the same value for all points in the same horizontal layer, provided that the air is in a state of equilibrium. On account of the great compressibility of gas, the lower layers are much more dense than the upper ones; but the density, like the pressure, is constant in value for the same horizontal layer, throughout any portion of air in a state of equilibrium. Whenever there is an inequality either of density or pressure at a given level, wind must ensue.

We owe to Torricelli an experiment which plainly shows the pressure of the atmosphere, and enables us to estimate its intensity with great precision. This experiment, which was performed in 1643, one year after the death of Galileo, at a time when the weight and pressure of the air were scarcely even suspected, has immortalized the name of its author, and has exercised a most important influence upon the progress of natural philosophy.

102. Torricelli's Experiment.—A tube of about a quarter or a third of an inch in diameter, and about a yard in length, is completely filled with mercury; the extremity is then stopped with the finger, and the tube is inverted in a vessel containing mercury. If the finger is now removed, the mercury will descend in the tube, and after a few

is true, it does not affect the numerical value of the weight when stated in grammes, but because the pressure of 760 millimetres of mercury varies as the intensity of gravity, so that more air is compressed into the space of a litre as gravity increases. (§ 107, 6.)

The *weight in grammes* is another name for the *mass*. The force of gravity on a litre of air under the pressure of 760 millimetres is proportional to the square of the intensity of gravity.

This is an excellent example of the ambiguity of the word *weight*, which sometimes denotes a mass, sometimes a force; and though the distinction is of no practical importance so long as we confine our attention to one locality, it cannot be neglected when different localities are compared.

Regnault's determination of the weight of a litre of dry air at 0° Cent. under the pressure of 760 millimetres at Paris is 1·293187 gramme. Gravity at Paris is to gravity at Greenwich as 3456 to 3457. The corresponding number for Greenwich is therefore 1·293561.

oscillations will remain stationary at a height which varies according to circumstances, but which is generally about 30 inches.

The column of mercury is maintained at this height by the pressure of the atmosphere upon the surfac⌐ of the mercury in the

Fig. 108.—Torricellian Experiment.

vessel. In fact, the pressure at the level ABCD must be the same within as without the tube; so that the column of mercury BE exerts a pressure equal to that of the atmosphere.

Accordingly, we conclude from this experiment of Torricelli that *every surface exposed to the atmosphere sustains a normal pressure equal, on an average, to the weight of a column of mercury whose base is this surface, and whose height is 30 inches.*

It is evident that if we performed a similar experiment with water, whose density is to that of mercury as 1 : 13·59, the height of the column sustained would be 13·59 times as much; that is,

30 × 13·59 inches, or about 34 feet. This is the maximum height to which water can be raised in a pump; as was observed by Galileo.

In general the heights of columns of different liquids equal in weight to a column of air on the same base, are inversely proportional to their densities.

103. Pressure of One Atmosphere.—We can easily calculate the amount of this pressure for a given surface, for example, a square inch. It is the weight of a column of mercury whose base is a square inch and height 30 inches, that is, the weight of 30 cubic inches of mercury; and as a cubic inch of mercury weighs about half a pound, the atmospheric pressure on a square inch is about 15 pounds. This pressure of 15 pounds[1] to the square inch is called the pressure *of one atmosphere;* it is exerted in a normal direction at all points on the surface of a body, and in consequence, as in the case of a body wholly immersed in a liquid, the resultant of the different elementary pressures is a vertical upward pressure equal to the weight of the air displaced. The effect of the air, therefore, is not,

Fig. 109.

as was formerly supposed, to press bodies to the surface of the earth; on the contrary, it tends to raise them, as in a liquid, but with comparatively small force, owing to its small density. It is upon this principle that the ascent of balloons depends, as we shall see hereafter.

104. Pascal's Experiments.—It is supposed, though without any decisive proof, that Torricelli derived from Galileo the definite conception of atmospheric pressure.[2] However this may be, when the experiment of the Italian philosopher became known in France in 1644, no one was capable of giving the correct explanation of it, and the famous

---

[1] As the weight of a cubic centimetre of mercury at zero is 13·596 grammes, the pressure of 760 millimetres is 13·596 × 76 = 1033·3 grammes per square centimetre = 14·70 pounds per square inch. 760 millimetres are 29·922 inches.

[2] In the fountains of the Grand-duke of Tuscany some pumps were required to raise water from a depth of from 40 to 50 feet. When these were worked, it was found that they would not draw. Galileo determined the height to which the water rose in their tubes, and found it to be about 32 feet; and as he had observed and proved that air has weight, he readily conceived that it was the weight of a column of the atmosphere which maintained the water at this height in the pumps. No very useful results, however, were expected from this discovery, until, at a later date, Torricelli adopted and greatly extended it. Desiring to repeat the experiment in a more convenient form, he conceived the idea of substituting for water a liquid that is 14 times as heavy, namely, mercury, rightly imagining that a column of one-fourteenth of the length would balance the force which sustained 32 feet of water (Biot, *Biographie Universelle*, article "Torricelli").—*D.*

doctrine that "nature abhors a vacuum," by which the rising of water in a pump was accounted for, was generally accepted. Pascal was the first to prove incontestably the falsity of this old doctrine, and to introduce a more rational belief. For this purpose he proposed or executed a series of ingenious experiments, and discussed minutely all the phenomena which were attributed to nature's abhorrence of a vacuum, showing that they were necessary consequences of the pressure of the atmosphere.

We may cite in particular the observation, made at his suggestion, that the height of the mercurial column decreases in proportion as we ascend. This beautiful and decisive experiment, which is repeated as often as heights are measured by the barometer, and which leaves no doubt as to the nature of the force which sustains the mercurial column, was performed for the first time at Clermont, and on the top of the Puy-de-Dôme, on the 19th September, 1648.

105. The Barometer.—By fixing the Torricellian tube in a permanent position, we have a means of measuring the amount of the atmospheric pressure at any moment; and this pressure may be expressed by the height of the column of mercury which it supports. Such an instrument is called a *barometer*. In order that its indications may be accurate, several precautions must be observed. In the first place, the liquid used in different barometers must be identical, for the height of the column supported naturally depends upon the density of the liquid employed, and if this varies, the observations made with different instruments will not be comparable.

The mercury employed is chemically pure, being generally made so by washing with a dilute acid and by subsequent distillation. The barometric tube is filled nearly full, and is then placed upon a sloping furnace, and heated till the mercury boils. The object of this process is to expel the air and moisture which may be contained in the mercurial column, and which, without this precaution, would gradually ascend

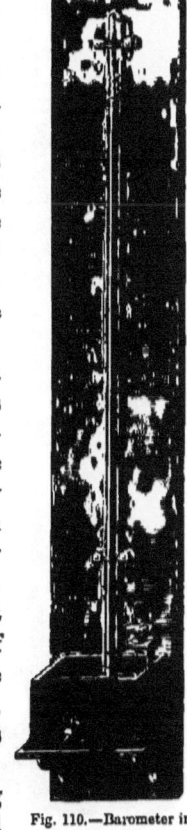

Fig. 110.—Barometer in its simplest form.

10

into the vacuum above, and cause a downward pressure of unknown amount, which would prevent the mercury from rising to the proper height.

The next step is to fill up the tube with pure mercury, taking care not to introduce any bubble of air. The tube is then inverted

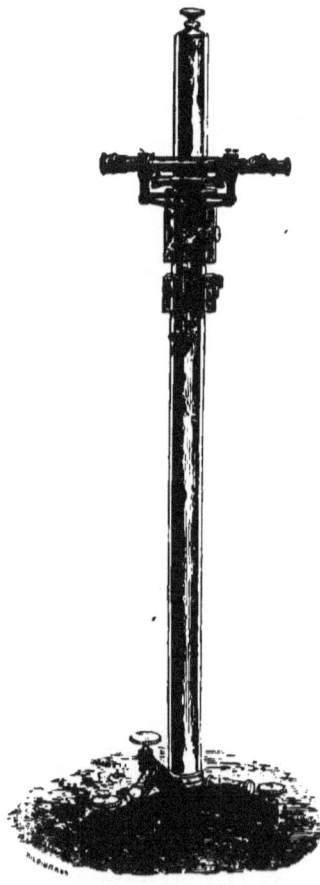

in a cistern likewise containing pure mercury recently boiled, and is firmly fixed in a vertical position, as shown in Fig. 110.

We have thus a fixed barometer; and in order to ascertain the atmospheric pressure at any moment, it is only necessary to measure the height of the top of the column of mercury above the surface of the mercury in the cistern. For this purpose an iron rod, working in a screw, is fixed vertically above the surface of the mercury in the dish. The extremities of this rod are pointed, and the lower extremity being brought down to touch the surface of the liquid below, the distance of the upper extremity from the top of the column of mercury is measured. Adding to this the length of the rod, which has previously been determined once for all, we have the barometric height. This measurement may be effected with great precision by means of the cathetometer.

105 A. Cathetometer.—This instrument, which is so frequently employed in physics to measure the vertical distance between two points, was invented by Dulong and Petit.

Fig. 111.—Cathetometer.

It consists essentially (Fig. 111) of a vertical scale divided usually into half millimetres. This scale forms part of a brass cylinder capable of turning very easily about a strong steel axis. This axis is fixed on a pedestal provided with three levelling screws, and with two spirit-levels at right angles to each other. Along the scale moves a sliding frame carrying a telescope furnished with crosswires, that is, with two very fine threads, usually spider lines, in the focus of the eye-piece, whose point of intersection serves to determine the line of vision. By means of a clamp and slow-motion screw, the telescope can be fixed with great precision at any required height. The telescope is also provided with a spirit-level and adjusting screw. When the apparatus is in correct adjustment, the line of vision of the telescope is horizontal, and the graduated scale is vertical. If then we wish to measure the difference of level between two points, we have only to sight them successively, and measure the distance passed over on the scale, which is done by means of a vernier attached to the sliding frame.

106. Fortin's Barometer.—The barometer just described is intended to be fixed; when portability is required the barometer invented by Fortin is employed. It is also perfectly adapted to general use. The cistern, which is formed of a tube of boxwood, surmounted by a tube of glass, is closed below by a piece of leather, which can be raised or lowered by means of a screw. This screw works in the bottom of a copper case, which incloses the cistern except at the middle, where it is cut away in front and at the back, so as to leave the surface of the mercury open to view. The barometric tube is encased in a tube of copper, with two slits at opposite sides (Fig. 113); and it is on this tube

Fig. 113.
Upper portion of Barometer.

Fig. 112.
Cistern of Fortin's Barometer.

that the divisions are engraved, the zero point from which they are reckoned being the lower extremity of an ivory point fixed in the covering of the cistern. The temperature of the mercury, which is required for one of the corrections mentioned in next section, is given by a thermometer with its bulb resting against the tube. A sliding

piece, furnished with a vernier,[1] moves along the tube by means of
the screw B, and enables us to determine the height with great precision.   Its lower edge is the zero of the vernier.   The way in which
the barometric tube is fixed upon the cistern is worth notice.   In the
centre of the upper surface of the copper casing there is an opening,
from which rises a short tube of the same metal, lined with a tube of
boxwood.   The barometric tube is push'ed inside, and fitted in with
a piece of chamois leather, which prevents the mercury from issuing;
but does not exclude the air, which, passing through the pores of the
leather, penetrates into the cistern, and so transmits its pressure.

Before taking an observation, the surface of the mercury is adjusted,
by means of the lower screw, to touch the ivory point.   The observer
knows when this condition is fulfilled by seeing the extremity of the
point touch its image in the mercury.   The vernier is then raised or
lowered, until the horizontal plane in which its zero lies is tangential

[1] The vernier is an instrument very largely employed for measuring the fractions of a
unit of length on any scale. Suppose we have a scale divided into inches, and another
scale containing nine inches divided into ten equal parts. If now we make the end of this
latter scale, which is called the vernier, coincide with one of the divisions in the scale of
inches, as each division of the vernier is $\frac{9}{10}$ of an inch, it is evident that the first division
on the scale will be $\frac{1}{10}$ of an inch beyond the first division on the vernier, the second on the
scale $\frac{2}{10}$ beyond the second on the vernier, and so on until the ninth on the scale, which

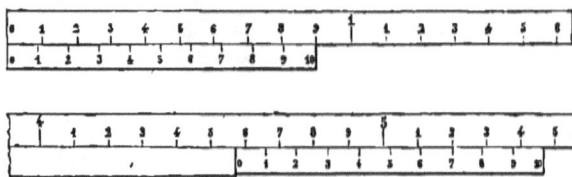

Fig. 114.—Vernier.

will exactly coincide with the tenth on the vernier. Suppose next that in measuring
any length we find that its extremity lies between the degrees 5 and 6 on the scale; we
bring the zero of the vernier opposite the extremity of the length to be measured, and
observe what division on the vernier coincides with one of the divisions on the scale. We
see in the figure that it is the seventh, and thus we conclude that the fraction required is
$\frac{7}{10}$ of an inch.

If the vernier consisted of 19 inches divided into 20 equal parts, it would read to the $\frac{1}{20}$
of an inch; but there is a limit to the precision that can thus be obtained. An exact coincidence of a division on the vernier with one on the scale seldom or never takes place, and
we merely take the division which approaches nearest to this coincidence; so that when
the difference between the degrees on the vernier and those on the scale is very small, there
may be so much uncertainty in this selection as to nullify the theoretical precision of the
instrument. Verniers are also employed to measure angles; when a circle is divided into
half degrees, a vernier is used which gives $\frac{1}{30}$ of a division on the circle, that is, $\frac{1}{60}$ of a
half degree, or one minute.—*D.*

to the upper surface of the mercurial column, as shown in Fig. 113. In making this adjustment, the back of the instrument should be turned towards a good light, in order that the observer may be certain of the position in which the light is just cut off at the summit of the convexity.

When the instrument is to be moved, the screw at the bottom is turned until the tube is filled. The cistern will then be full also, and the barometer should be inverted, as an additional safeguard both against the introduction of air and the escape of mercury. In making observations upon the surface of the ground, the instrument is suspended from a tripod stand by gimbals,[1] so that it always takes a vertical position; or it may be fixed permanently against a wall.

106A. Float Adjustment.—In some barometers the ivory point for indicating the proper level of the mercury in the cistern is replaced by a float. F (Fig. 107)

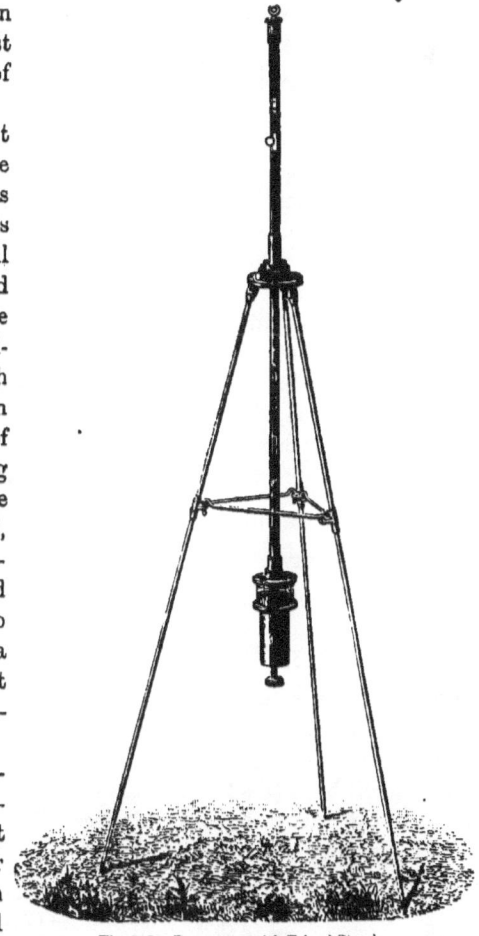

Fig. 115.—Barometer with Tripod Stand.

is a small ivory piston, having the float attached to its foot, and moving freely up and down between the two ivory guides I. A horizontal line (interrupted by the piston) is engraved on the two

---

[1] A kind of universal joint, in common use on board ship for the suspension of compasses, lamps, &c. It is seen in Fig. 115, at the top of the tripod stand.

guides, and another is engraved on the piston, at such a height that the three lines form one straight line when the surface of the mercury in the cistern stands at the zero point of the scale.

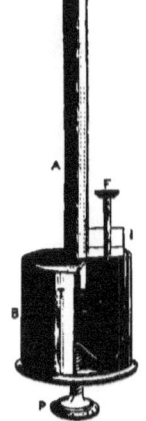

Fig. 107.
Float Adjustment.

107. Barometric Corrections. — In order that barometric heights may be comparable as measures of atmospheric pressure, certain corrections must be applied.

1. *Correction for Temperature.* As mercury expands with heat, it follows that a column of warm mercury exerts less pressure than a column of the same height at a lower temperature; and it is usual to reduce the actual height of the column to the height of a column at the temperature of freezing water, which would exert the same pressure.

Let h be the observed height at temperature $t°$ Centigrade, and h_o the height reduced to freezing-point. Then, if m be the coefficient of expansion of mercury per degree Cent., we have

$$h_o (1 + m\ t) = h, \text{ whence } h_o = h - h\ m\ t \text{ nearly.}$$

The value of m (Chap. xxii.) is $\frac{1}{5550} = ·00018018$. For temperatures Fahrenheit, we have

$$h_o \left\{ 1 + m\ (t - 32) \right\} = h, \ h_o = h - h\ m\ (t - 32),$$

where m denotes $\frac{1}{9990} = ·0001001$.

But temperature also affects the length of the divisions on the scale by which the height of the mercurial column is measured. If these divisions be true inches at $0°$ Cent., then at $t°$ the length of n divisions will be $n\ (1 + l\ t)$ inches, l denoting the coefficient of linear expansion of the scale, the value of which for brass, the usual material, is ·00001878. If then the observed height h amounts to n divisions of the scale, we have

$$h_o (1 + m\ t) = h = n\ (1 + l\ t);$$

whence

$$h_o = \frac{n\ (1 + l\ t)}{1 + m\ t} = n - n\ t\ (m - l), \text{ nearly};$$

that is to say, if n be the height read off on the scale, it must be diminished by the correction $n\ t\ (m - l)$, t denoting the temperature of the mercury in degrees Centigrade. The value of $m - l$ is ·0001614.

For temperatures Fahrenheit, assuming the scale to be of the correct length at 32° Fahr., the formula for the correction (which is still subtractive), is n $(t-32)$ $(m-l)$, where $m-l$ has the value ·00008967.[1]

2. *Correction for Capillarity.*—In the preceding chapter we have seen that mercury in a glass tube undergoes a capillary depression, whence it follows that the observed barometric height is too small, and that we must add to it the amount of this depression. In all tubes of internal diameter less than about ½ of an inch this correction is sensible; and its amount, for which no simple formula can be given, has been computed, from theoretical considerations, for various sizes of tube, by several eminent mathematicians, and recorded in tables, from which that given below is abridged. These values are applicable on the assumption that the meniscus which forms the summit of the mercurial column is decidedly convex, as it always is when the mercury is rising. When the meniscus is too flat, the mercury must be lowered by the foot-screw, and then screwed up again.

It is found by experiment, that the amount of capillary depression is only half as great when the mercury has been boiled in the tube, as when this precaution has been neglected.

For purposes of special accuracy, tables have been computed,

---

[1] The correction for temperature is usually made by the help of tables, which give its amount for all ordinary temperatures and heights. These tables, when intended for English barometers, are generally constructed on the assumption that the scale is of the correct length not at 32° Fahr., but at 62° Fahr., which is (by act of Parliament) the temperature at which the British standard yard (preserved in the office of the Exchequer) is correct. On this supposition, the length of n divisions of the scale at temperature $t°$ Fahr., is

$$n\left\{1+l\,(t-62)\right\};$$

and by equating this expression to

$$h_o\left\{1+m\,(t-32)\right\}$$

we find

$$h_o=n\left\{1-m\,(t-32)+l\,(t-62)\right\}$$
$$=n\left\{1-(m-l)\,t+(32\,m-62\,l)\right\}$$
$$=n\left\{1-·00008967\,t+·00255654\right\};$$

which, omitting superfluous decimals, may conveniently be put in the form—

$$n-\frac{n}{1000}\,(·09\,t-2·56).$$

The correction vanishes when

$$·09\,t-2·56=0;$$

that is, when $t=\dfrac{256}{9}=28·5$.

For all temperatures higher than this the correction is subtractive.

giving the amount of capillary depression for different degrees of convexity, as determined by the sagitta (or height) of the meniscus, taken in conjunction with the diameter of the tube. Such tables, however, are seldom used in this country.[1]

TABLE OF CAPILLARY DEPRESSIONS IN UNBOILED TUBES.

(To be halved for Boiled Tubes.)

Diameter of tube in inches.	Depression.	Diameter.	Depression.	Diameter.	Depression.
·10	·140	·20	·058	·40	·015
·11	·126	·22	·050	·42	·013
·12	·114	·24	·044	·44	·011
·13	·104	·26	·038	·46	·009
·14	·094	·28	·033	·48	·008
·15	·086	·30	·029	·50	·007
·16	·079	·32	·026	·55	·005
·17	·073	·34	·023	·60	·004
·18	·068	·36	·020	·65	·003
·19	·063	·38	·017	·70	·002

3. *Correction for Capacity.*—When there is no provision for adjusting the level of the mercury in the cistern to the zero point of the scale, another correction must be applied. It is called the correction for *capacity*. In barometers of this construction, which were formerly much more common than they are at present, there is a certain point in the scale at which the mercurial column stands when the mercury in the cistern is at the correct level. This is called the neutral point. If A be the interior area of the tube, and C the area of the cistern (exclusive of the space occupied by the tube and its contents), when the mercury in the tube rises by the amount x, the mercury in the cistern falls by an amount $y = \frac{A}{C}x$, for the volume of the mercury which has passed from the cistern into the tube is $Cy = Ax$. The change of atmospheric pressure is correctly measured by $x + y = \left(1 + \frac{A}{C}\right)x$, and if we now take x to denote the distance of the summit of the mercurial column from the neutral point, the corrected distance will be $\left(1 + \frac{A}{C}\right)x$, and the correction to be applied to the observed reading will be $\frac{A}{C}x$, which is additive if the observed reading be above the neutral point, subtractive if below.

It is worthy of remark that the neutral point depends upon the

---

[1] The most complete collection of meteorological and physical tables, is that edited by Professor Guyot, and published under the auspices of the Smithsonian Institution, Washington.

volume of mercury. It will be altered if any mercury be lost or added; and as temperature affects the volume, a special temperature-correction must be applied to barometers of this class. The investigation will be found in a paper by Professor Swan in the *Philosophical Magazine* for 1861.

In some modern instruments the correction for capacity is avoided, by making the divisions on the scale less than true inches, in the ratio $\frac{C}{A+C}$, and the effect of capillarity is at the same time compensated by lowering the zero point of the scale. Such instruments, if correctly made, simply require to be corrected for temperature.

4. *Index Errors.*—Under this name are included errors of graduation, and errors in the position of the zero of the graduations. An error of zero makes all readings too high or too low by the same amount. Errors of graduation (which are generally exceedingly small) are different for different parts of the scale.

Barometers intended for accurate observation are now usually examined at Kew Observatory before being sent out; and a table is furnished with each, showing its index error at every half inch of the scale, errors of capillarity and capacity (if any) being included as part of the index error. We may make a remark here once for all respecting the signs attached to errors and corrections. The sign of an error is always opposite to that of its correction. When a reading is too high the index error is one of excess, and is therefore positive; whereas the correction needed to make the reading true is subtractive, and is therefore negative.

5. *Reduction to Sea-level.*—In comparing barometric observations taken over an extensive district for meteorological purposes, it is usual to apply a correction for difference of level. Atmospheric pressure, as we have seen, diminishes as we ascend; and it is usual to add to the observed height the difference of pressure due to the elevation of the place above sea-level. The amount of this correction is proportional to the observed pressure. The law according to which it increases with the height will be discussed in the next chapter.

6. *Correction for Unequal Intensity of Gravity.*—When two barometers indicate the same height, at places where the intensity of gravity is different (for example, at the pole and the equator), the same mass of air is superincumbent over both; but the pressures are unequal, being proportional to the intensity of gravity as measured by the values of g (Chaps. v. vi.) at the two places. When intensity

of pressure is to be expressed in absolute measure, it should be stated in absolute units of force (§ 42) per unit area. If we adopt as our absolute unit of force, that force which, acting on a pound of matter for a second, would generate a velocity of a foot per second, it is necessary that the square foot should be made the unit of area.

Since the force of gravity on a pound contains g absolute units of force, and the weight of 144 cubic inches of mercury at 0° Centigrade is 70·7275 lbs., we have the following rules for reducing pressure per unit area to absolute measure:—

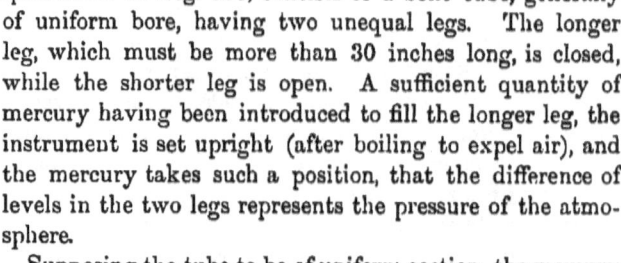

To reduce lbs. per sq. foot to absolute measure, multiply by g.

   ,,   lbs. per sq. inch   ,,   ,, $144\,g$.

   ,, -   inches of mercury   ,,   ,, $g \times 70·7275$.

108. Other kinds of Mercurial Barometer.—The *Siphon Barometer*, which is represented in Fig. 116, consists of a bent tube, generally

of uniform bore, having two unequal legs. The longer leg, which must be more than 30 inches long, is closed, while the shorter leg is open. A sufficient quantity of mercury having been introduced to fill the longer leg, the instrument is set upright (after boiling to expel air), and the mercury takes such a position, that the difference of levels in the two legs represents the pressure of the atmosphere.

Supposing the tube to be of uniform section, the mercury will always fall as much in one leg as it rises in the other. Each end of the mercurial column therefore rises or falls through only half the height corresponding to the change of atmospheric pressure.

In the best siphon barometers there are two scales, one for each leg, as indicated in the figure, the divisions on one being reckoned upwards, and on the other downwards,

Fig. 116.
Siphon
Barometer.

from an intermediate zero point, so that the sum of the two readings is the difference of levels of the mercury in the two branches.

Inasmuch as capillarity tends to depress both extremities of the mercurial column, its effect is generally neglected in siphon barometers; but practically it causes great difficulty in obtaining accurate observations, for according as the mercury is rising or falling its extremity is more or less convex, and a great deal of tapping is usually required to make both ends of the column assume the same form, which is the condition necessary for annihilating the effect of capillary action.

Wheel Barometer.—The wheel barometer, which is in more general use than its merits deserve, consists of a siphon barometer, the two branches of which have usually the same diameter. On the surface of the mercury of the open branch floats a small piece

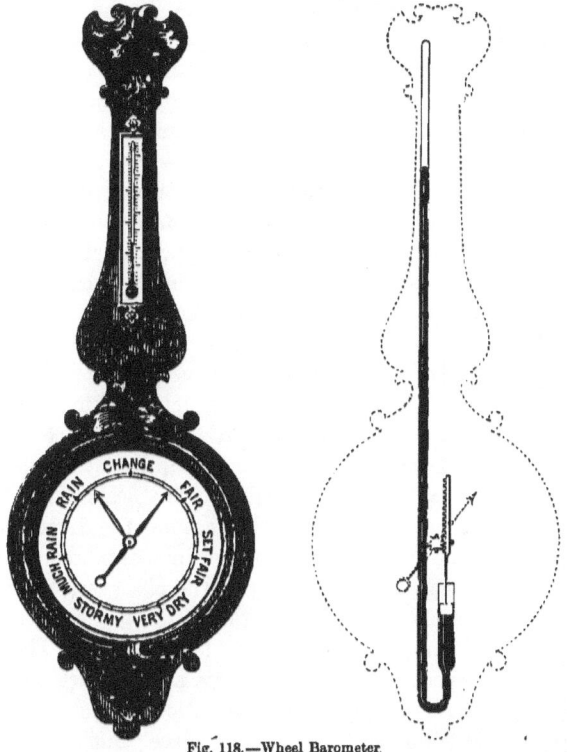

Fig. 118.—Wheel Barometer.

of iron or glass suspended by a thread, the other extremity of which is fixed to a pulley, on which the thread is partly rolled. Another thread, rolled parallel to the first, supports a weight which balances the float. To the axis of the pulley is fixed a needle which moves on a dial. When the level of the mercury varies in either direction, the float follows its movement through the same distance; by the action of the counterpoise the pulley turns, and with it the needle, the extremity of which points to the figures on the dial, marking the baro-

metric heights. The mounting of the dial is usually placed in front
of the tube, so as to conceal its presence. The wheel barometer is a
very old invention, and was introduced by the celebrated Hooke in
1683. The pulley and strings are sometimes replaced by a rack and
pinion, as represented in the figure (Fig. 118).

Besides the faults incidental to the siphon barometer, the wheel
barometer is encumbered in its movements by the friction of the
additional apparatus. It is quite unsuitable for measuring the exact
amount of atmospheric pressure, and is slow in indicating changes.

Marine Barometer.—The ordinary mercurial barometer cannot be
used at sea, on account of the violent oscillations which the mercury
would experience from the motion of the vessel. In order to meet

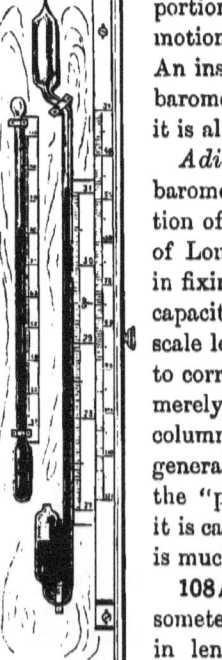

this difficulty, the tube is contracted in its middle
portion nearly to capillary dimensions, so that the
motion of the mercury in either direction is hindered.
An instrument thus constructed is called a marine
barometer. When such an instrument is used on land
it is always too slow in its indications.

Adie's Barometer.—A very convenient form of
barometer, which is extensively used under the direction of the Board of Trade, is constructed by Adie
of London. The error of capillarity is allowed for
in fixing the zero point of the scale. The error of
capacity is obviated by making the divisions of the
scale less than true inches, in such a ratio as exactly
to correct for capacity. The observer, therefore, has
merely to read the height of the top of the mercurial
column, and correct for temperature. The tube is
generally contracted in its middle part, to diminish
the "pumping" (*i.e.* oscillation), which occurs when
it is carried from place to place; but the contraction
is much less than in the marine barometer.

108 A. Sympiesometer (συν, πιεζω).—Adie's sympiesometer (Fig. 117) consists of a glass tube 18 inches
in length and ¾ inch in diameter, with a small
chamber at the top, and an open cistern below. In
the original construction the upper part of the tube
was filled with hydrogen, and the lower part and
cistern with oil of almonds. In the construction now employed these
materials are replaced by common air and glycerine.

Fig. 117
Sympiesometer.

When the pressure of the atmosphere increases, the air in the upper part of the tube is compressed, and the fluid rises; when it diminishes, the fluid falls. The instrument is graduated by comparison with a mercurial barometer. The intervals corresponding to inches of mercurial pressure are much longer than inches, and are of unequal length, becoming shorter as we ascend on the tube. To obviate error from the increased pressure of the inclosed air when its temperature is raised, a thermometer and sliding scale are added to the instrument, so that it may be adjusted for temperature at each observation. The sympiesometer is very quick in its indications, and from its portability is well adapted for being used at sea, but it is not suited for exact observation.

As originally made it was liable to gradual change, from absorption of the hydrogen by the oil of almonds. In the present construction absorption is less liable to occur, at least if the glycerine be of the proper consistency.

109. Aneroid Barometer (α, νηρος).—This barometer depends upon

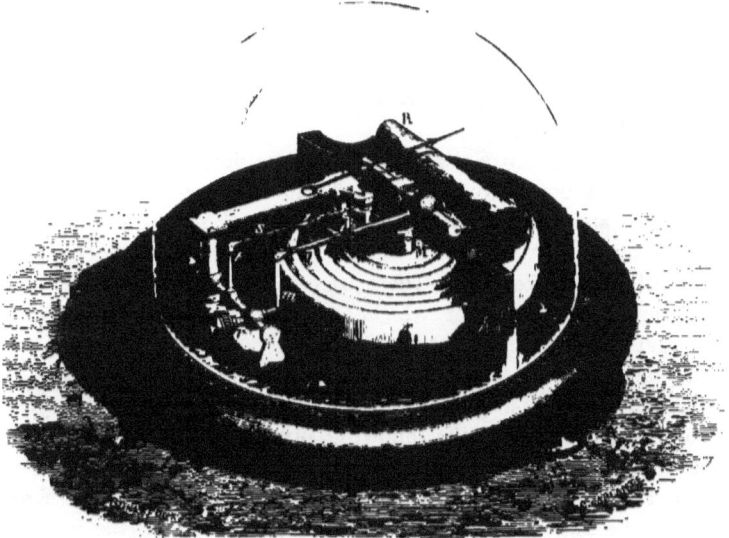

Fig. 119.—Aneroid Barometer.

the changes in the form of a thin metallic vessel, partially exhausted of air, as the atmospheric pressure varies. M. Vidie was the first to overcome the numerous difficulties which were presented in the construction of these instruments. , We subjoin a figure of the model which he finally adopted.

The essential part is a cylindrical box partially exhausted of air, the upper surface of which is corrugated in order to make it yield more easily to external pressure. At the centre of the top of the box is a small metallic pillar M, which acts upon a powerful steel spring R. As the pressure varies, the top of the box rises or falls, transmitting its movement to the spring, and thence, by means of two levers l and m, to a metallic axis r. This latter carries a third lever t, the extremity of which is attached to a chain s which turns a drum, the axis of which bears the index needle. A spiral spring keeps the chain constantly stretched, and thus makes the needle always take a position corresponding to the shape of the box at the time. The graduation is performed empirically by comparison with a mercurial barometer. The aneroid barometer is very sensitive, and is much more portable than any form of mercurial barometer, being both lighter and less liable to injury. It is sometimes made small enough for the waistcoat pocket. It has the drawback of being affected by temperature to an extent which must be determined for each instrument separately, and of being liable to gradual changes which can only be checked by occasional comparison with a good mercurial barometer.

In the *metallic barometer*, which is a modification of the aneroid, the exhausted box is crescent-shaped, and the horns of the crescent separate or approach according as the external pressure diminishes or increases.

110. Old Forms Revived.—There are two ingenious modifications of the form of the barometer, which, after long neglect, have recently been revived for special purposes.

Counterpoised Barometer.—The invention of this instrument is attributed to Samuel Morland, who constructed it about the year 1680. It depends upon the following principle:—If the barometric tube is suspended from one of the scales of a balance, there will be required to balance it in the other scale a weight equal to the weight of the tube and the mercury contained in it, minus the upward pressure of the liquid against the bottom of the tube and its contents.[1]

---

[1] It may be shown that if a be the (annular) area of a section of the tube itself, and A the area of the inclosed space (which is filled with mercury), the resultant force to which the tube is subjected from atmospheric and liquid pressure combined is a downward force

$$PA - pa,$$

If the atmospheric pressure increases, the mercury will rise in the tube, and consequently the weight of the floating body will increase, while the upward pressure will be slightly diminished on account of the sinking of the mercury in the cistern. The beam will thus incline to the side of the baro metric tube, and the reverse would be the case if the pressure diminished. For the balance may be substituted, as in the figure, a lever carrying a counterpoise; the variations of pressure will be indicated by the movements of this lever.

Such an instrument may very well be used as a *barograph* or recording barometer; for this purpose we have only to attach to the lever an arm with a pencil, which is con-

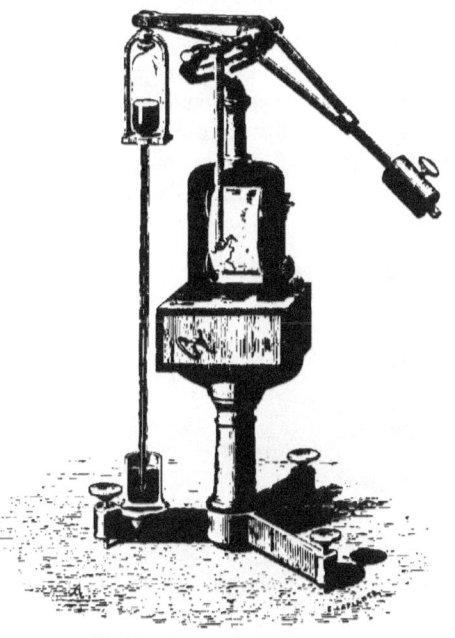

Fig. 120.—Counterpoised Barometer.

stantly in contact with a sheet of paper moved uniformly by clockwork. The result will be a continuous trace, whose form corresponds to the variations of pressure. It is very easy to determine, either by calculation or by comparison with a standard barometer, the pressure corresponding to a given position of the pencil on the paper;

P denoting atmospheric pressure, and p the fluid pressure due to the depth of immersion (exclusive of the transmitted atmospheric pressure). This resultant force together with the weight of the tube must be equal to the supporting force at the point of suspension. If the latter be constant, $PA - pa$ must be constant, and the changes in P and p must be inversely as the areas A and a. If these areas are equal P and p will be equal; that is, the tube will descend through the same distance as the mercury in a common barometer would rise; and if A is greater than a, the movement will be proportionately magnified. For great sensitiveness, therefore, the tube should be large and thin.

We have here neglected the changes of level in the mercury in which the tube is immersed. These changes tend to increase the distance moved by the tube, and must be added to the movements as above calculated.

thus, if the paper is ruled with twenty-four equidistant lines, corresponding to the twenty-four hours of the day, we can see at a glance what was the pressure at any given time. An arrangement of this kind has been adopted by the Abbé Secchi for the meteorograph of the observatory at Rome. The first successful employment of this kind of barograph appears to be due to Mr. Alfred King, a gas engineer of Liverpool, who invented and constructed such an instrument in 1853, for the use of the Liverpool Observatory, and subsequently designed a larger one, which is still in use, furnishing a very perfect record, magnified five-and-a-half times.

Fahrenheit's Barometer.—Fahrenheit's barometer consists of a tube

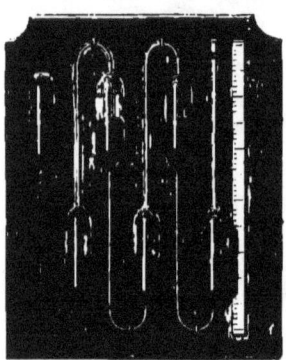

Fig. 121.—Fahrenheit's Barometer.

bent several times, the lower portions of which contain mercury; the upper portions are filled with water, or any other liquid, usually coloured. It is evident that the atmospheric pressure is balanced by the sum of the differences of level of the columns of mercury, diminished by the sum of the corresponding differences for the columns of water; whence it follows that, by employing a considerable number of tubes, we may greatly reduce the height of the barometric column. This circumstance renders the instrument interesting as a scientific curiosity, but at the same time diminishes its sensitiveness, and renders it unfit for purposes of precision. It is therefore never used for the measurement of atmospheric pressure; but an instrument upon the same principle has recently been employed for the measurement of very high pressures, as will be explained in Chap. xiv.

110 A. **Photographic Registration.**—Since the year 1847 various meteorological instruments at the Royal Observatory, Greenwich, have been made to yield continuous traces of their indications by the aid of photography, and the method is now generally employed at meteorological observatories in this country. The Greenwich system is fully described in the *Greenwich Magnetical and Meteorological Observations* for 1847, pp. lxiii.–xc. (published in 1849).

The general principle adopted for all the instruments is the same. The photographic paper is wrapped round a glass cylinder, and the axis of the cylinder is made parallel to the direction of the movement which is to be registered. The cylinder is turned by clockwork, with uniform velocity. The spot of light (for the magnets and barometer), or the boundary of the line of light (for the thermometers), moves, with the movements which are to be registered, backwards and forwards in the direction of the axis of the cylinder, while the cylinder itself is turned round. Consequently (as in Morin's machine, Chap. v.), when the paper is unwrapped from its cylindrical form, there is traced upon it a curve of which the abscissa is proportional to the time, while the ordinate is proportional to the movement which is the subject of measure.

The barometer employed in connection with this system is a large siphon barometer, the bore of the upper and lower extremities of its arms being about 1·1 inch. A glass float in the quicksilver of the lower extremity is partially supported by a counterpoise acting on a light lever (which turns on delicate pivots), so that the wire supporting the float is constantly stretched, leaving a definite part of the weight of the float to be supported by the quicksilver. This lever is lengthened to carry a vertical plate of opaque mica with a small aperture, whose distance from the fulcrum is eight times the distance of the point of attachment of the float-wire, and whose movement, therefore (§ 108), is four times the movement of the column of a cistern barometer. Through this hole the light of a lamp, collected by a cylindrical lens, shines upon the photographic paper.

Every part of the cylinder, except that on which the spot of light falls, is covered with a case of blackened zinc, having a slit parallel to the axis of the cylinder; and by means of a second lamp shining through a small fixed aperture, and a second cylindrical lens, a base line is traced upon the paper, which serves for reference in subsequent measurements.

The whole apparatus, or any other apparatus which serves to give a continuous trace of barometric indications, is called a *barograph;* and the names *thermograph, magnetograph, anemograph,* &c., are similarly applied to other instruments for automatic registration.

11

CHAPTER XIII.

111. Measurement of Heights by the Barometer.—As the height of the barometric column diminishes when we ascend in the atmosphere, it is natural to seek in this phenomenon a means of measuring heights. The problem would be extremely simple, if the air had everywhere the same density as at the surface of the earth. In fact, the density of the air at sea-level being about 10,500 times less than that of mercury, it follows that, on the hypothesis of uniform density, the mercurial column would fall an inch for every 10,500 inches, or 875 feet, that we ascend. This result, however, is far from being in exact accordance with fact, inasmuch as the density of the air diminishes very rapidly as we ascend, on account of its great compressibility.

111 A. Height of Homogeneous Atmosphere.—If the atmosphere were of uniform and constant density, its height would be approximately obtained by multiplying 30 inches by 10,500, which gives 26,250 feet, or about 5 miles.

More accurately, if we denote by H the height of the atmosphere at a given time and place, on the assumption that the density throughout is the same as the observed density D at the base, and if we denote by P the observed pressure at the base, expressed in absolute units of force per unit area (§ 107, 6), then since the pressure P must be equal to the weight of a column of volume H and of mass HD, we have

$$P = g \, HD \qquad H = \frac{P}{g \, D} \qquad (1)$$

The height H, computed on this imaginary assumption, is called the *height of the homogeneous atmosphere*, corresponding to the pressure P, density D, and intensity of gravity g, and is frequently introduced in physical formulæ.

The expression for H shows that its value is not affected if P and D vary in the same ratio, as is the case in barometric fluctuations when the temperature is constant; but that increase of temperature and increase of moisture increase H, since warm air and moist air are less dense than cold and dry air at the same pressure.

It is not necessary that.the height H should be reckoned from the surface of the earth. It may be reckoned upwards from any point in the atmosphere, and denotes the height which the air above this point would have, if reduced to the density D which exists at the point.

Neglecting differences of temperature and moisture, and the trifling diminution of gravity as we ascend, the value of H is the same for all points in the same vertical column, because, as we ascend, P and D diminish in the same ratio.

112. Principles of Hypsometry.—Supposing the temperature, moisture, and intensity of gravity to be uniform in a vertical column of air, it is easy to state the law according to which the pressure would diminish as we ascend. Consider, for example, three layers of equal thickness, which is so small that we may regard the density as constant within the limits of each layer, though varying from each layer to the next. Let D, D′, D″ be their densities, and P, P′, P″ the pressures at their lower faces, the weights of the two lower layers are $P-P'$ and $P'-P''$,

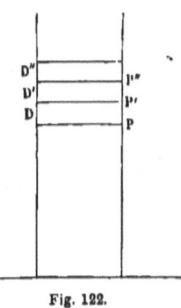

Fig. 122.

and these must be proportional to their densities; hence we have

$$\frac{P-P'}{P'-P''} = \frac{D}{D'}$$

but by Boyle's law $\frac{D}{D'} = \frac{P}{P'}$; consequently we have $\frac{P-P'}{P'-P''} = \frac{P}{P'}$, whence it easily follows that $\frac{P}{P'} = \frac{P'}{P''}$; that is to say, the ratio of the density of the first layer to that of the second, is the same as of the second to the third. Applying this principle to any number of consecutive layers of equal thickness, we see that the ratio of the density of each to that of the next will be the same for the whole series. It follows that, *as the heights increase in arithmetical progression, the pressures diminish in geometrical progression.*

This proposition may be put into the algebraical form:

$$x_2 - x_1 = H \log \frac{P_1}{P_2},$$

where x_1, x_2 are the heights of two stations above a fixed level, P_1, P_2 the pressures at the two stations, and H some constant. The proof given in the note[1] shows, that if the logarithms are Napierian, H is equal to the height of the homogeneous atmosphere. If the logarithms be of the common kind, H is equal to the height of the homogeneous atmosphere multiplied by 2·3026, the value of which product for the latitude of Great Britain, and for the temperature of freezing water, is about 60,360 feet.

This formula has been deduced on the supposition that the temperature and the intensity of gravity are uniform through the whole extent of the air between the two stations. If these two elements vary, they cause the value of H to vary; and it would be necessary for accuracy to employ in the formula the mean value of H for the stratum of air which intervenes between the two stations. The variation in the intensity of gravity is usually insignificant, and it is customary to assume as the mean temperature, the arithmetical mean of the temperatures t_1 and t_2 of the two stations. On these assumptions the value of H, if the temperatures be expressed in degrees Fahrenheit, will (by the law of expansion of air, Chap. xxiii.) be

$$60,360 \left(1 + \frac{t_1 + t_2 - 64}{986}\right).$$

It is proved in treatises on logarithms that if $\frac{P_1}{P_2}$ be but little greater than unity,

$$\text{Nap. log } \frac{P_1}{P_2} = 2 \frac{P_1 - P_2}{P_1 + P_2} \text{ nearly;}$$

and since the height of the homogeneous atmosphere at freezing temperature in these latitudes is about 26,214 feet, we obtain the formula—

$$\text{Difference of level in feet} = 52428 \frac{P_1 - P_2}{P_1 + P_2} \left(1 + \frac{t_1 + t_2 - 64}{986}\right),$$

which may be used for differences of level not exceeding about 3000 feet.

[1] Let x denote distance measured upwards from a fixed level, then, using the notation of § 111A, the pressure due to the weight of a layer of thickness dx is $g\,D\,dx$; but this is the amount by which the pressure diminishes as x is increased by the amount dx; we have therefore

$$-d\,P = g\,D\,dx = \frac{P}{H}dx,$$

since by § 111A, $g\,D = \frac{P}{H}$. We have therefore

$$-\frac{d\,P}{P} = \frac{dx}{H}, \text{ whence log } P_1 - \text{log } P_2 = \frac{x_2 - x_1}{H}.$$

The determination of heights by means of atmospheric pressure, whether the pressure be observed directly by the barometer or indirectly by the boiling-point thermometer (see Chap. xxvi.), is called *hypsometry* (ὕψος, height).

As a rough rule, it may be stated that, in ordinary circumstances, the barometer falls an inch in ascending 900 feet.

113. Diurnal Oscillation of the Barometer.—In these latitudes, the mercurial column is in a continual state of irregular oscillation; but in the tropics it rises and falls with great regularity according to the hour of the day, attaining two maxima in the twenty-four hours.

It generally rises from 4 A.M. to 10 A.M., when it attains its first maximum; it then falls till 4 P.M., when it attains its first minimum; a second maximum is observed at 10 P.M., and a second minimum at 4 A.M. The hours of maxima and minima are called the tropical hours (τρεπω, to turn), and vary a little with the season of the year. The difference between the highest maximum and lowest minimum is called the diurnal[1] *range*, and the half of this is called the *amplitude* of the diurnal oscillation. The amount of the former does not exceed about a tenth of an inch.

The character of this diurnal oscillation is represented in Fig. 123. The vertical lines correspond to the hours of the day; lengths have been measured upwards upon them proportional to the barometric heights at the respective hours, diminished by a constant quantity; and the points thus determined have been connected by a continuous curve. It will be observed that the two lower curves, one of which relates to Cumana, a town of Venezuela, situated in about 10° north latitude, show strongly marked oscillations corresponding to the maxima and minima. In our own country the regular diurnal oscillation is marked by irregular fluctuations, so that a single day's observations give no clue to its existence. Nevertheless, on taking observations at regular hours for a number of consecutive days, and comparing the mean heights for the different hours, some indications of the law will be found. A month's observations will be sufficient

---

[1] The epithets *annual* and *diurnal*, when prefixed to the words *variation, range, amplitude*, denote the *period* of the variation in question; that is, the time of a complete oscillation. Diurnal variation does not denote variation from one day to another, but the variation which goes through its cycle of values in one day of twenty-four hours. Annual range denotes the range that occurs within a year. This rule is universally observed by writers of high scientific authority.

A table, exhibiting the values of an element for each month in the year, is a table of annual (not monthly) variation; or it may be more particularly described as a table of variations from month to month.

for an approximate indication of the law; but observations, extending over some years, will be required to establish with anything like

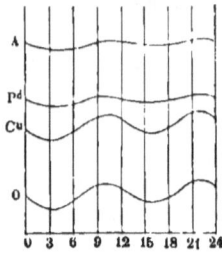

Fig. 123.
Curves of Diurnal Variation.

precision the hours of maxima and the amplitude of the oscillation.

The two upper curves represent the diurnal variation of the barometer at Padua (lat. 45° 24′) and Abo (lat. 60° 56′), the data having been extracted from Kaemtz's *Meteorology.* We see, by inspection of the figure, that the oscillation in question becomes less strongly marked as the latitude increases. The range at Abo is less than half a millimetre. At about the 70th degree of north latitude it becomes insensible; and in approaching still nearer to the pole, it appears from observations, which however need further confirmation, that the oscillation is reversed; that is to say, that the maxima here are contemporaneous with the minima in lower latitudes.

There can be little doubt that the diurnal oscillation of the barometer is in some way attributable to the heat received from the sun, which produces expansion of the air, both directly, as a mere consequence of heating, and indirectly, by promoting evaporation, and thus increasing the volume of the air (as well as diminishing its sp. gravity) by the addition of aqueous vapour. The precise nature of the connection between this cause and the diurnal barometric oscillation has not, however, as yet been satisfactorily established.

114. Irregular Variations of the Barometer.—The height of the barometer, at least in the temperate zones, depends on the state of the atmosphere, and its variations often serve to predict the changes of weather with more or less certainty. In this country the barometer generally falls for rain or S.W. wind, and rises for fine weather or N.E. wind.

Barometers for popular use have generally the words—

Set fair.	Fair.	Change.	Rain.	Much rain.	Stormy.
30·5	30	29·5	29	28·5	28 inches.

marked at the respective heights. These words must not, however, be understood as absolute predictions. A low barometer rising is generally a sign of fine, and a high barometer falling of wet weather. Moreover, it is to be borne in mind that the barometer stands about a tenth of an inch lower for every hundred feet that we ascend above sea-level.

The connection between a low or falling barometer and wet weather is to be found in the fact that moist air is specifically lighter than dry, even at the same temperature, and still more when, as usually happens, moist air is warmer than dry.

115. Inverse March of Barometer and Thermometer.—It is impossible to lay down universal rules for the connection between the indications of the barometer and the state of the weather, since rules which would usually hold true in one place might be quite inapplicable at another. We may, however, state a principle which is of very extensive application, namely, that warm winds, especially when they have passed over considerable masses of water, are likely to be accompanied by rain; for they are charged with vapour which is liable to be condensed as its temperature falls. Cold winds, on the contrary, contain vapour which was taken up at a lower temperature than it now has, and is therefore far from a state of saturation. They are therefore unlikely to produce rain unless it be when they first begin to blow, when they may condense vapour previously existing in the air.

These characteristics are very marked in our own country, where the warm winds from the south-west have passed over the Atlantic, while the cold winds from the north-east have for the most part traversed dry land.

Again, the march of the barometer is in general opposite to that of the thermometer; that is to say, *the barometer usually falls when the thermometer rises,* and *vice versâ.* This law is one of the most general in meteorology, and is easily explained; in fact, when the temperature rises at any place, it produces a dilatation of the air, and consequently an overflow into neighbouring regions; the weight of air over the place is thus diminished. On the contrary, a fall of temperature produces an inflow of air and an increase of pressure.

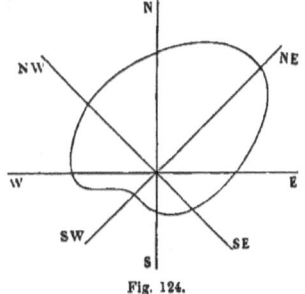

Fig. 124.
Barometric Rose of Winds at Paris.

It is therefore to be expected that the mean barometric height should be lower during warm and rainy winds than during cold and dry winds; and that this is the case is rendered extremely evident by the annexed figure, which represents the mean barometric rose of the winds at Paris. Upon each of the eight lines which represent the principal directions of the wind, have been laid off lengths proportional to the corresponding barometric pressures diminished by a constant. The figure shows a very sudden increase in passing from S.W. to W. It is, in fact, during the change of the wind from one of these quarters to the other, that the greatest atmospheric perturbations occur.

116. **Synoptic Weather Charts. Isobaric Lines.**—The extension of telegraphic communication over Europe has led to the establishment of a system of correspondence by which the barometric pressures, at a given moment, at a number of stations which have been selected for meteorological observation, are known at one or more stations appointed for receiving the reports. From the information thus furnished, curves (called isobaric lines) are drawn upon a chart through those places at which the pressure is the same. The barometric condition of an extensive region is thus rendered intelligible at a glance. Plate I. is a specimen of these synoptic charts,[1] which are prepared every day at the observatory of Paris; it refers to the 22d of January, 1868. Besides the isobaric lines, these charts indicate, by the system of notation explained at the left of the figure, the general state of the weather, the strength of wind, and state of the sea. The isobaric curves correspond to differences of five millimetres (about 0·2 inch) of pressure, and according as they are near together, or far apart, the variation of pressure in passing from one to another is more or less sudden (or to use a very expressive modern phrase, the barometric gradient is more or less steep), just as the contour lines on a map of hilly ground approach each other most nearly where the ground is steepest. Generally speaking, the wind blows from regions of high to regions of low barometer, and with greater force as the barometric gradient is steeper.

The isobaric lines frequently, as in the example here selected, form closed curves encircling a region of barometric depression. Two such centres are here exhibited—one in the south of England and the other in the west of Russia. Such centres of depression always accompany great atmospheric disturbances. The air, in fact, rushes

---

[1] The curves drawn upon this chart are isobaric lines, each corresponding to a particular barometric pressure, which is indicated by the numerals marked against it. These denote the pressure in millimetres diminished by 700. For example, the line which passes through the south of Spain corresponds to the pressure 770 millimetres; that through the north of Spain to 765 millimetres. The curves are drawn for every fifth millimetre. The smaller numerals, which are given to one place of decimals, indicate the pressures actually observed at the different stations, from which the isobaric lines are drawn by estimation.

The other symbols refer to cloud, wind, and sea, and are explained at the left of the chart.

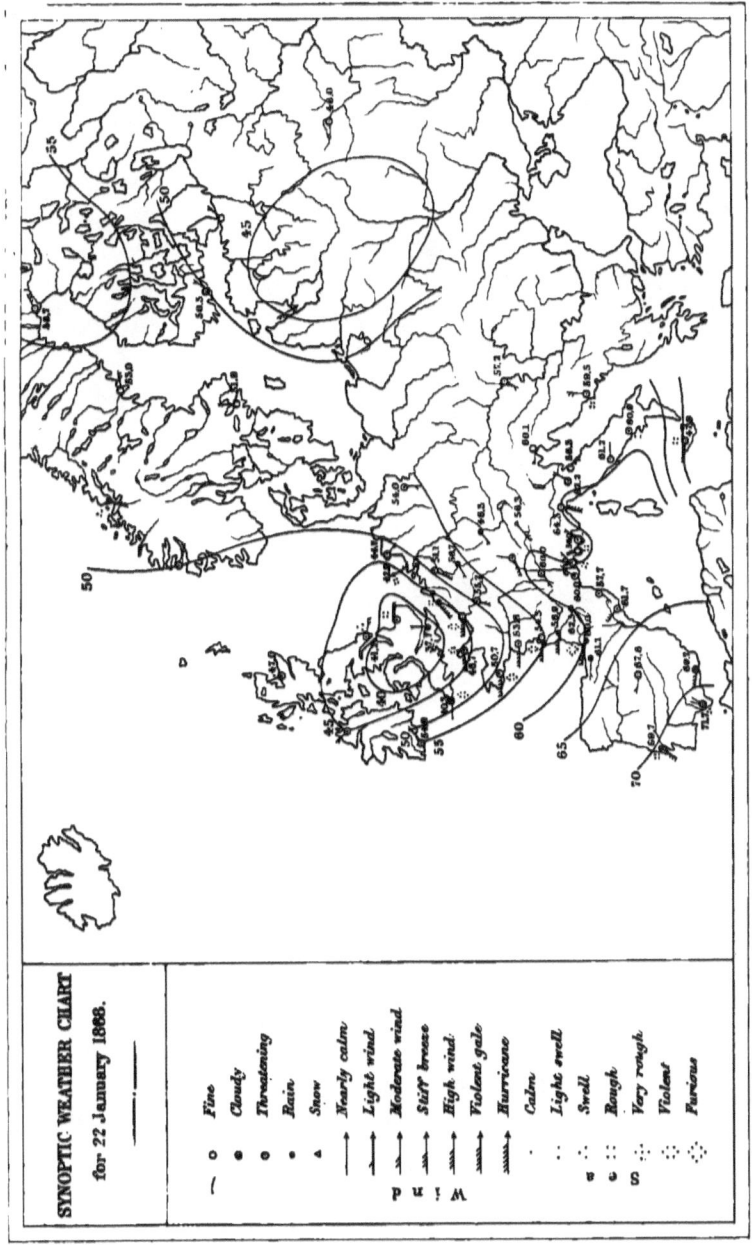

in from all sides, usually with a spiral motion, towards these centres of depression, the direction of rotation in the spiral being, for the northern hemisphere, opposite to the motion of the hands of a watch with its face upwards. The centrifugal force due to this rotation tends to increase the central depression, and thus protracts the duration of the phenomenon.

These revolving storms are called cyclones. They attain their greatest violence in tropical regions, the West Indies being especially noted for their destructive effect. They frequently proceed from the Gulf of Mexico in a north-easterly direction, increasing in diameter as they proceed, but diminishing in violence. Their velocity of translation is usually from ten to twenty miles an hour.

The storm-warnings inaugurated by the late Admiral Fitzroy are based partly upon information received by telegraph of storms that have actually commenced at some distant locality, and partly upon a comparison of barometric pressures at different localities.[1]

---

[1] For fuller information respecting the laws of storms, which is a purely modern subject, and is continually receiving fresh developments, we would refer to Mr. Buchan's *Handy Book of Meteorology*. See also § 406 A in Part II. of the present Work.

It will be observed, by the arrows in the annexed chart, that the direction of the wind, instead of being coincident with the line of steepest descent from each isobaric curve to the next below it, generally makes a large angle (considerably exceeding 45°) to the right of it. This law (known as Buys Ballot's) is general for the northern hemisphere, and is dependent on the earth's rotation (§ 406 A). The influence of the earth's rotation in modifying the direction of winds, is discussed in a paper "On the General Circulation and Distribution of the Atmosphere," by the Editor of this Work, in the *Philosophical Magazine* for September, 1871.

CHAPTER XIV.

BOYLE'S (OR MARIOTTE'S) LAW.[1]

117. Boyle's Law.—As gases are composed of molecules in a state of permanent repulsion, they may be compared to springs constantly bent, and making constant efforts to free themselves. The amount of pressure which they exert against the sides of the vessels which contain them, depends upon the volume which they occupy, increasing as this volume diminishes. By a number of careful experiments upon this point, Boyle and Mariotte independently established the law that this volume varies inversely as the pressure, provided that the temperature remain constant. As the density evidently varies inversely as the volume, we may express the law in other words by saying that at the same temperature the density varies directly as the pressure.

If V and V' be the volumes of the same quantity of gas, P and P', D and D', the corresponding pressures and densities, Boyle's law will be expressed by the equations

$$\frac{V}{V'} = \frac{P'}{P} = \frac{D'}{D}.$$

118. Mariotte's Tube.—The correctness of this law may be verified by means of the following apparatus, which was employed by both the experimenters above named. It consists (Fig. 125) of a bent tube with branches of unequal length; the long branch is open, and the

---

[1] Boyle, in his *Defence of the Doctrine touching the Spring and Weight of the Air against the Objections of Franciscus Linus*, appended to *New Experiments, Physico-mechanical, &c.* (second edition, 4to, Oxford, 1662), describes the two kinds of apparatus represented in Figs. 125, 126 as having been employed by him, and gives in tabular form the lengths of tube occupied by a body of air at various pressures. These observed lengths he compares with the theoretical lengths computed on the assumption that volume varies reciprocally as pressure, and points out that they agree within the limits of experimental error.

Mariotte's treatise, *De la Nature de l'Air*, is stated in the *Biographie Universelle* to have been published in 1679. (See Preface to Tait's *Thermodynamics*, p. iv.)

short branch closed. The tube is fastened to a board provided with two scales; one by the side of the long branch, divided into parts of equal length; the other by the side of the short branch, having divisions which correspond to parts of equal volume. The graduation of both scales begins from the same horizontal line through 0, 0. Mercury is first poured in at the extremity of the long branch, and by inclining the apparatus to either side, and cautiously adding more of the liquid if required, the mercury can be made to stand at the same level in both branches, and at the zero of both scales. Thus we have, in the short branch, a quantity of air separated from the external air, and at the same pressure. Mercury is then poured into the long branch, so as to reduce the volume of this inclosed air by one-half; it will then be found that the difference of level of the mercury in the two branches is equal to the height of the barometer at the time of the experiment; the compressed air therefore exerts a pressure equal to that of two atmospheres. If more mercury be poured in so as to reduce the volume of the air to one-third or one-fourth of the original volume, it will be found that the difference of level is respectively two or three times the height of the barometer; that is, that the compressed air exerts a pressure equal respectively to that of three or four atmospheres. This experiment therefore shows that if the volume of the gas becomes two, three, four times as small, the pressure becomes two, three, four times as great. This is the principle expressed in Boyle's law.

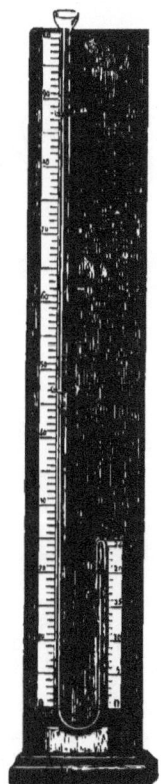

Fig. 125.
Mariotte's Tube.

The law may also be verified in the case where the gas expands, and where its pressure consequently diminishes. For this purpose a barometric tube (Fig. 126), partially filled with mercury, is inverted in a tall vessel, containing mercury also, and is held in such a position that the level of the liquid is the same in the tube and in the vessel. The volume occupied by the gas is marked, and the tube is raised; the gas expands, its pressure diminishes, and, in virtue of the excess of the atmospheric pressure, a column of mercury ab rises in the tube, so that its height, added to the pressure of the expanded air, is

equal to the atmospheric pressure.   It will then be seen that if the
volume of air becomes double what it was before, the height of the
column raised is one-half that of the barometer;
that is, the expanded air exerts a pressure equal
to half that of the atmosphere.   If the volume is
trebled, the height of the column is two-thirds
that of the barometer; that is, the pressure of the
expanded air is one-third that of the atmosphere,
a result which is in accordance with Boyle's law.

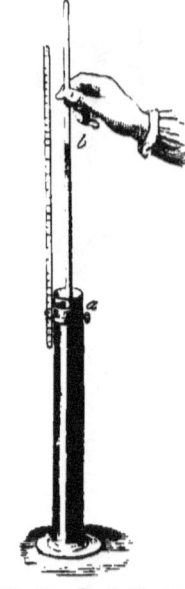

Fig. 120.—Proof of Boyle's
Law for Expanding Air.

119. Despretz's Experiments.—The simplicity of
Boyle's law, taken in conjunction with its apparent agreement with facts, led to its general accep-
tance as a rigorous truth of nature, until in 1825
Despretz published an account of experiments,
showing that different gases are unequally compressible.   He inverted in a cistern of mercury
several cylindrical tubes of equal height, and filled
them with different gases.   The whole apparatus
was then inclosed in a strong glass vessel filled
with water, and having a screw-piston as in
Œrsted's piesometer (§ 22).   On pressure being
applied, the mercury rose to unequal heights in
the different tubes, carbonic acid for example
being more reduced in volume than air.   These
experiments proved that though Boyle's law might possibly be true
for one of the gases employed, it could not be rigorously true for
more than one.

In 1829 Dulong and Arago undertook a laborious series of experiments with the view of testing the accuracy of the law as applied to
air; and the results which they obtained, even when the pressure was
increased to twenty-seven atmospheres, agreed so nearly with it as
to confirm them in the conviction that, for air at least, it was rigorously true.   When re-examined, in the light of later researches, the
results obtained by Dulong and Arago seem to point to a different
conclusion.

120. Unequal Compressibility of Different Gases.—The unequal compressibility of different gases, which was first established by Despretz's
experiments above described, is now usually exhibited by the aid of the
following apparatus designed by Pouillet.   A is a cast-iron reservoir,
containing mercury surmounted by oil.   In this latter liquid dips a

bronze plunger P, the upper part of which has a thread cut upon it, and works in a nut, so that the plunger can be screwed up or down by means of the lever L. The reservoir A communicates by an iron tube with another cast-iron vessel, into which are firmly fastened two tubes T T about six feet in length and $\frac{1}{16}$th of an inch in internal diameter, very carefully calibrated (§ 180). Equal volumes of two gases, perfectly dry, are introduced into these tubes through their upper ends, which are then hermetically sealed. The plunger is then made to descend, and a gradually increasing pressure is exerted, the volumes occupied by the gases are measured, and it is ascertained that no two gases follow precisely the same law of compression. The difference, however, is almost insensible when the gases employed are non-liquefiable, as air, oxygen, hydrogen, nitrogen, nitric oxide, and marsh-gas. But when we compare any one of these with a liquefiable gas, such as carbonic acid, cyanogen, or ammonia, the difference is rapidly and distinctly manifested. Thus, under a pressure of twenty-five atmospheres, carbonic acid occupies a volume which is only $\frac{4}{5}$ths of that occupied by air.

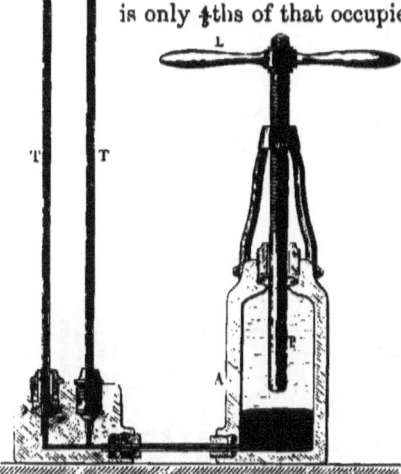

Fig. 127.—Pouillet's Apparatus for showing Unequal Compressibility of Different Gases.

121. Regnault's Experiments.—Boyle's law, therefore, is not to be considered as rigorously exact; but it is certainly a very close approximation to the truth, except for gases near their point of liquefaction. In order to demonstrate the inaccuracy of the law for air, or any gas that is not liquefiable, and more especially if it is required to determine the law of deviation for each particular gas, it is necessary to employ very precise methods of measurement. In ordinary experiments on compression, and even in the elaborate investigations of Dulong and Arago, a definite portion of gas is taken

and successively diminished in volume by the application of continually increasing pressure. Now it is evident that, in
experiments of this kind, in proportion as the pressure
increases, the variations in volume become smaller, and
the precision with which they can be determined consequently diminishes. Regnault adopted the plan of
operating in all cases upon the same volume of gas,
which, being initially at different pressures, was always
reduced to one-half. The pressure was observed before
and after this operation, and, if Boyle's law were true, its
value should be found to be doubled. In this way the
same precision of measurement is obtained at high as at
low pressures.

A general view of Regnault's apparatus is given in Fig.
128. It consists of an iron reservoir containing mercury,
furnished at the top with a
force-pump for water. The
lower part of this reservoir communicates with a cylinder which
is also of iron, and in which are
two openings to admit tubes.
Communication between the
reservoir and the cylinder can
be established or interrupted by
means of a stop-cock R, of very
exact workmanship. Into one

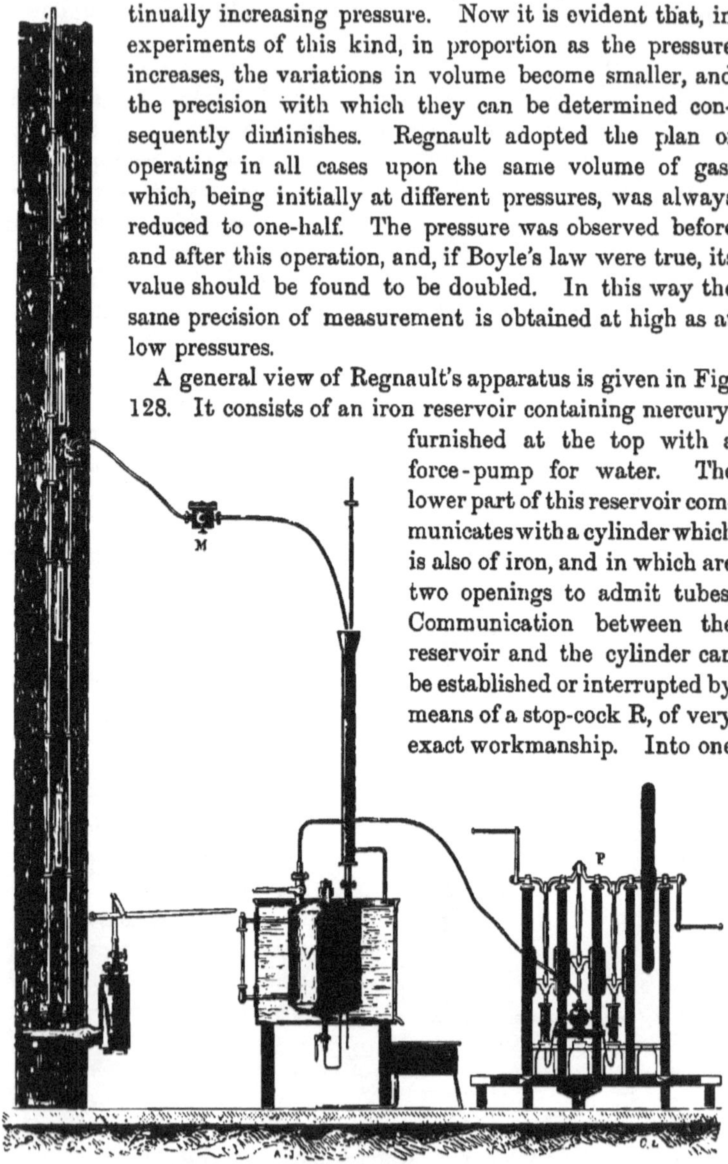

Fig. 128.—Regnault's Apparatus for Testing Boyle's Law.

of the openings is fitted the lowest of a series of glass tubes A, which are placed end to end, and firmly joined to each other by metal fittings, so as to form a vertical column of about twenty-five metres in height.

The height of the mercurial column in this long manometric tube could be exactly determined by means of reference marks placed at distances of about ·95 of a metre, and by the graduation on the tubes forming the upper part of the column. The mean temperature of the mercurial column was given by thermometers placed at different heights. Into the second opening in the cylinder fits the lower extremity of the tube B, which is divided into millimetres, and also gauged with great accuracy. This tube has at its upper end a stop-cock r which can open communication with the reservoir V, into which the gas to be operated on is forced and compressed by means of the pump P.

An outer tube, which is not shown in the figure, envelops the tube B, and, being kept full of water, which is continually renewed, enables the operator to maintain the tube at a temperature sensibly constant, which is indicated by a very delicate thermometer. Before fixing the tube in its place, the point corresponding to the middle of its volume is carefully ascertained, and after the tube has been permanently fixed, the distance of this point from the nearest of the reference marks is observed.[1]

After these explanatory remarks we may describe the mode of conducting the experiments. The gas to be operated on, after being first thoroughly dried, is introduced at the upper part of the tube B, the stop-cock of the pump being kept open, so as to enable the gas to expel the mercury and occupy the entire length of the tube. The force-pump is then brought into play, and the gas is reduced to about half of its former volume; the pressure in both cases being ascertained by observing the height of the mercury in the long tube above the nearest mark. It is important to remark that it is not at all necessary to operate always upon exactly the same initial volume, and reduce it exactly to one-half, which would be a very tedious opera-

---

[1] Regnault's apparatus was fixed in a small square tower of about fifteen metres in height, forming part of the buildings of the Collége de France, and which had formerly been built by Savart for experiments in hydraulics. The tower could therefore contain only the lower part of the manometric column; the upper part rose above the platform at the top of the tower, resting against a sort of mast which could be ascended by the observer. The readings inside the tower could be made by means of a cathetometer, but this was impossible in the upper portion of the column, and for this reason the tubes forming this portion were graduated.—D.

tion; these two conditions are approximately fulfilled, and the graduation of the tube enables the observer always to ascertain the actual volumes.

122. Results.—The general result of the investigations of Regnault is, that Boyle's law does not exactly represent the compressibility of even non-liquefiable gases, such as air, hydrogen, nitrogen, which, with carbonic acid, were the gases operated on by him. But the resulting differences are so small that they would not be detected by a mere inspection of the numbers which represent volumes and pressures. They may, however, be clearly exhibited by submitting the results to the following test:—Suppose we take a certain quantity of gas which, under the pressure P, occupies the volume V, and that we reduce it to a volume V', when the pressure becomes P', and, if Boyle's law were accurate, we should have the equation

$$\frac{V}{V'} = \frac{P'}{P}, \text{ or } VP = V'P'$$

Or, in another form,

$$\frac{VP}{V'P'} - 1 = 0.$$

Now it is found that for all gases except hydrogen this difference, instead of being always zero, is constantly positive, and has not only a sensible value, but, which is of especial importance, it increases regularly with the pressure, which shows that it cannot be attributed to the inevitable errors of observation.

If we measure off upon any line lengths proportional to the different pressures, and raise perpendiculars proportional to the differences $\frac{VP}{V'P'} - 1$, by joining the extremities of these perpendiculars by a continuous line an uninterrupted curve is obtained, which evidently is a graphic representation of the departure of the gas in question from Boyle's law. These curves have been very carefully traced by Regnault; and their algebraic equations can be found by the ordinary methods of interpolation. These equations are employed when it is required to calculate rigorously the change of volume corresponding to a very high pressure.

Since the difference $\frac{VP}{V'P'} - 1$ is positive, V'P' must be less than VP, and consequently the volume V' corresponding to the pressure P' is less than that given by Boyle's law. We thus see that, in general, gases are more compressible than Boyle's law would indicate; and in the case of gases that are liquefiable, this difference of compressibility is, as we have said, considerable.

In this respect hydrogen is a remarkable exception, as was originally shown by Despretz in the experiments which we have mentioned; it is less compressible than it should be by Boyle's law, the difference $\frac{VP}{V'P'} - 1$ being negative.

This singular peculiarity of hydrogen is quite in harmony with the views which are entertained as to the nature of this gas. It has been observed, from several comparative experiments performed upon carbonic acid, that at the temperature of 100° Cent. the law of compressibility of this gas differs from Boyle's law much less than at ordinary temperatures. We may thus fairly suppose that if we were to operate at a still higher temperature, we should approach still more nearly to the law, which would doubtless be verified at a particular temperature, beyond which the error would be in the opposite direction. It would thus appear that for each gas there is a sort of normal temperature, at which the compressibility is exactly represented by Boyle's law.

The compressibility of a gas should also increase as the temperature decreases, as is proved by the experiment on carbonic acid. With the exception of hydrogen, all gases under ordinary conditions are below this normal temperature. But if, as chemical phenomena tend to prove, hydrogen is a kind of metal, we must suppose it to be relatively in a high state of rarefaction, which accounts for the peculiarity presented by its compressibility.

123. Manometers or Pressure-gauges.—Manometers are instruments for measuring the elastic force of a gas or vapour contained in the interior of a closed space. This elastic force is generally expressed in units called atmospheres (§ 103), and is often measured by means of a column of mercury.

When the column of mercury moves freely in an open tube, the manometer is said to be open [à air libre]; it was a manometer of this kind that Regnault employed to measure the successive pressures to which the volume of gas was subjected.

The open mercurial pressure-gauge is often used in the arts to measure pressures that are not very considerable. The figure represents one of the simplest forms. The apparatus consists of a box, generally of iron, at the top of which is an opening closed by a screw stopper, which is traversed by the tube b, open at both ends, and dipping into the mercury in the box. The air or vapour whose elastic force is to be measured enters by the tube a, and presses upon

12

the mercury.   It is evident that if the level of the liquid in the box
is the same as in the tube, the pressure in the box must be exactly
equal to that of the atmosphere.   If the mercury in the tube rises
above that in the box, the pressure of the air in the box must exceed
that of the atmosphere by a pressure corresponding to the height of
the column raised.   The pressures are generally marked in atmospheres upon a scale beside the tube.

124. Multiple Branch Manometer.—When the pressures to be measured
are considerable, as in the boiler of a high-pressure steam-engine, the

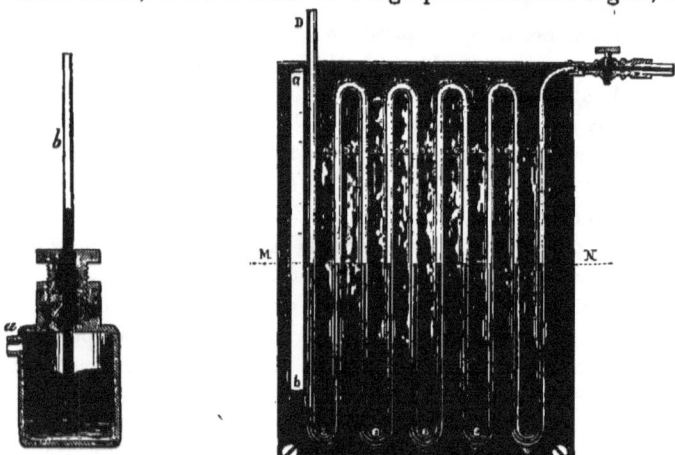

Fig. 129.—Open
Mercurial Manometer.   Fig. 130.—Multiple Branch Manometer.

above instrument, if employed at all, must be of a length corresponding to the pressure.   If, for instance, the pressure in question is eight
atmospheres, the length of the tube must be at least 8×30 inches=
20 feet.   Such an arrangement is inconvenient even for stationary
machines, and is entirely inapplicable to movable machines.

Without departing from the principle of the open mercurial pressure-gauge, namely, the balancing of the pressure to be observed
against the weight of a liquid increased by one atmosphere, we may
reduce the length of the instrument by an artifice already employed
by Fahrenheit in his barometer (§110).

The apparatus for this purpose consists of an iron tube ABCD bent
back upon itself a certain number of times.   The extremity A communicates with the boiler by a stop-cock, and the last branch CD is
of glass, and has a scale by its side.

The first step is to fill the tube with mercury as far as the level MN. At this height are holes by which the mercury escapes when it reaches them, and which are afterwards hermetically sealed. The upper portions are filled with water through openings which are also stopped after the tube has been filled. If the mercury in the first tube, which is in communication with the reservoir of gas, falls through a certain distance h, it will alternately fall and rise through the same distance in each of the tubes, and will consequently rise through the same distance in the last tube; now this distance corresponds to an effective pressure represented by a column of mercury of height $10h$, diminished by ten times the same height of water; that is, to a height of mercury equal to $10h \left(1 - \frac{1}{13 \cdot 59}\right)$. It will thus be seen that a very considerable pressure will be indicated by a comparatively small variation of the mercurial column. If, for instance, beginning with the atmospheric pressure, an additional pressure of five atmospheres is exerted, that is, an effective pressure of six atmospheres, the quantity h will be given in metres by the equation

$$5 \times \cdot 76 = 10h \left(1 - \frac{1}{13 \cdot 59}\right),$$

whence

$$5 \times \cdot 76 \times 13 \cdot 59 = 10h \times 12 \cdot 59,$$
$$h = \cdot 40m.$$

125. Compressed-air Manometer.—This instrument, which may assume different forms, sometimes consists, as in Fig. 131, of a bent tube AB closed at one end a, and containing within the space Aa a quantity of air, which is cut off from external communication by a column of mercury. The apparatus has been so constructed, that when the pressure on B is equal to that of the atmosphere, the mercury stands at the same height in both branches; so that, under these circumstances, the inclosed air is exactly at atmospheric pressure. But if the pressure increases, the mercury is forced into the left branch, so that the air in that branch is compressed, and its tension gradually increases until

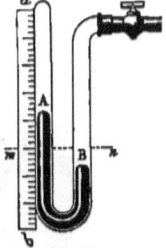

Fig. 131.—Compressed-air Manometer.

equilibrium is established. The pressure of the gas exerted at B is then equal to the pressure of the compressed air, together with that of a column of mercury equal to the difference of level of the liquid in the two branches. This pressure is expressed in atmospheres on the scale ab.

The graduation of this scale is effected directly in practice, by placing the manometer in communication with a reservoir of compressed air whose pressure is given by an open mercurial gauge, or by a standard manometer of any kind.

If the tube AB be supposed cylindrical, the graduation can be previously effected by an application of Boyle's law.

Let l be the length of the tube occupied by the inclosed air when its pressure is equal to that of one atmosphere; at the point to which the level of the mercury rises is marked the number 1. It is required to find to what point the end of the liquid column should reach when a pressure of n atmospheres is exerted at B. Let x be the height of this point above 1; then the volume of the air, which was originally l, has become $l-x$, and its pressure is therefore equal to $H \frac{l}{l-x}$, H being the mean height of the barometer. This pressure, together with that due to the difference of level $2x$, is equivalent to n atmospheres. We have thus the equation—

$$H \frac{l}{l-x} + 2x = nH,$$

whence

$$2x^2 - (nH + 2l)x + (n-1)Hl = 0.$$

$$x = \frac{nH + 2l \pm \sqrt{(nH + 2l)^2 - 8(n-1)Hl}}{4}.$$

We thus find two values of x; but that given by taking the positive sign of the radical is inadmissible; for if we put $n=1$, we ought to have $x=o$, which cannot be the case unless the sign of the radical is negative.

Fig. 132.
Compressed-air Manometer.

By giving n the successive values of $1\frac{1}{2}$, 2, $2\frac{1}{2}$, 3, &c., in this expression for x, we have the points on the scale corresponding to pressures of one atmosphere and a half, two atmospheres, &c.

As we have before remarked, the sensibility of the instrument decreases as the pressure increases, and the distance traversed by the mercury for an increment of pressure equal to one atmosphere becomes less and less. This inconvenience is partly avoided by the arrangement shown in Fig. 132. The branch containing the air is of a conical form; in this way, as the mercury rises, equal changes of volume correspond to increasing lengths. The effect of this arrangement is seen by an inspection of the scale, on which the numbers corresponding to successive atmospheres of pressure are nearly equidistant, whereas when the tube is cylindrical they rapidly approach each other.

126. Metallic Manometers.—The fragility of glass tubes, and the fact that they are liable to become opaque by the mercury clinging to their sides, are serious drawbacks to their use, especially in machines in motion. Accordingly, metallic manometers are often employed, depending upon the changes of form effected by the pressure of gas on its containing vessel, when suitably constructed. We shall here mention only Bourdon's gauge (Fig. 133). It consists essentially of a copper tube of elliptic section, which is bent through about 540° as represented in Fig. 133. One of the extremities communicates by a stop-cock with the reservoir of steam or compressed gas; to the other extremity is attached a steel needle which traverses a scale. When the stop-cock permits communication with the atmosphere, the end of the needle stands at the mark 1; but if the pressure increases the curvature diminishes, the free extremity of the tube moves away from the fixed extremity, and the needle traverses the scale.

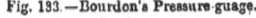

Fig. 133.—Bourdon's Pressure-guage.

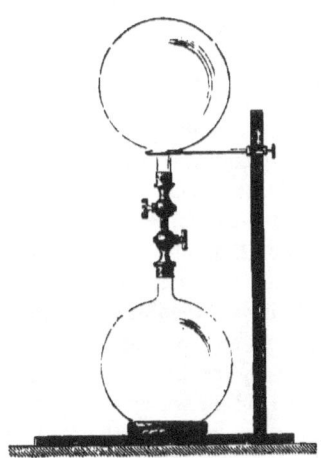

Fig. 134.—Mixture of Gases.

127. Mixture of Gases.—When gases of different densities are inclosed in the same space, experiment shows that, even under the most unfavourable circumstances, an intimate mixture takes place, so that each gas becomes uniformly diffused through the entire space. This fact has been shown by a decisive experiment due to Berthollet. He took two globes (Fig. 134) which could be screwed together, and placed them in a cellar. The lower globe was filled with carbonic acid, the upper globe with hydrogen. Communication was established between them, and at the end of a certain time it was ascertained that the gases had become intimately mixed; in fact, the proportion of carbonic acid and of hydrogen was exactly the same in both globes. The fact that the composition of the air is the same at all heights is another striking proof.

If several gases are inclosed in the same space, each of them exerts the same pressure as if the others were absent, and consequently the pressure exerted by the mixture is equal to the sum of the pressures due to each gas separately. These separate pressures can easily be calculated by Boyle's law, when the original pressure and volume of each gas are known.

For example, let V and P, V' and P', V'' and P'' be the volumes and pressures of the gases which are made to pass into a vessel of volume U. The first gas exerts, when in this vessel, a pressure equal to $\frac{VP}{U}$, the second a pressure equal to $\frac{V'P'}{U}$, the third a pressure equal to $\frac{V''P''}{U}$, and so on, so that the total pressure M is equal to $\frac{VP}{U} + \frac{V'P'}{U} + \frac{V''P''}{U}$, whence $MU = VP + V'P' + V''P''$.

This formula expresses the law of pressure for a mixture of gases; it may easily be verified by passing different volumes of gas into a graduated glass jar inverted over mercury, after having first measured their volumes and pressures.

128. Absorption of Gases by Liquids and Solids.—All gases are to a greater or less extent soluble in water. This property is of considerable importance in the economy of nature; thus the life of aquatic animals and plants is sustained by the oxygen of the air which the water holds in solution. The *volume* of a given gas that can be dissolved in water at a given temperature is found to be in general the same at all pressures,[1] and the ratio of this volume to that of the water which dissolves it is called the *coefficient of solubility or of absorption.* At the temperature 0° Cent. the coefficient of solubility for carbonic acid is 1, for oxygen ·04, and for ammonia 1150.

If a mixture of two or more gases be placed in contact with water,

---

[1] Hence the *weight* of gas absorbed is directly as the pressure.

each gas will be dissolved to the same extent as if it were the only gas present.

Other liquids as well as water possess the power of absorbing gases, according to the same laws, but with coefficients of solubility which are different for each liquid.

Increase of temperature diminishes the coefficient of solubility, which is reduced to zero when the liquid boils.

Some solids, especially charcoal, possess the power of absorbing gases. Boxwood charcoal absorbs about nine times its volume of oxygen, and about ninety times its volume of ammonia. When saturated with one gas, if put into a different gas, it gives up a portion of that which it first absorbed, and takes up in its place a quantity of the second. Finely-divided platinum condenses on the surface of its particles a large quantity of many gases, amounting in the case of oxygen to many times its own volume. If a jet of hydrogen gas be allowed to fall, in air, upon a ball of spongy platinum, the gas combines rapidly, in the pores of the metal, with the oxygen of the air, giving out an amount of heat which renders the platinum incandescent and usually sets fire to the jet of hydrogen.

Most solids have in ordinary circumstances a film of air adhering to their surfaces. Hence iron filings, if carefully sprinkled on water, will not be wetted, but will float on the surface, and hence also the power which many insects have of running on the surface of water without wetting their feet. The film of air in these cases prevents wetting, and hence, by the principles of capillarity, produces increased buoyancy.

CHAPTER XV.

129. Air-pump.—The air-pump was invented by Otto Guericke about 1650, and has since undergone some improvements in detail which have not altered the essential parts of its construction.

It consists of a glass or metal cylinder called the barrel, in which a piston works. This piston has an opening through it which is closed at the lower end by a valve S opening upwards. The barrel communicates with a passage leading to the centre of a brass surface carefully polished, which is called the plate of the air-pump. The entrance to the passage is closed by a conical stopper S', at the extremity of a metal rod which passes through the piston-head, and works in it tightly, so as to be carried up and down with the motion

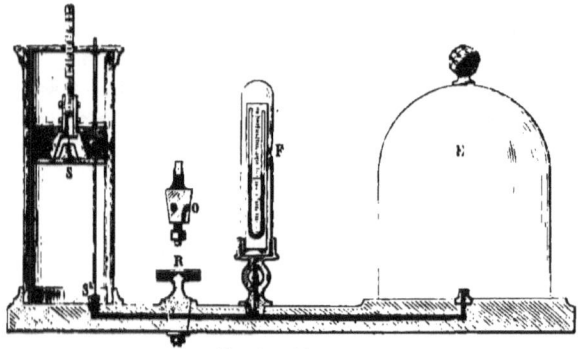

Fig. 135.—Air-pump.

of the piston. A catch at the upper part of the rod confines its motion within very narrow limits, and only permits the stopper to rise a small distance above the opening.

Suppose now that the piston is at the bottom of the cylinder, and is raised. The valve S' is opened, and the air of the receiver E rushes into the cylinder. On lowering the piston, the valve S' closes its opening, the air which has entered the cylinder cannot return into the receiver, and, on being compressed, raises the valve in the piston, and escapes into the air outside. On raising the piston again, a portion of the air remaining in the receiver will pass into the cylinder, whence it will escape on pushing down the piston, and so on.

We see, then, that if this motion be continued, a fresh portion of the air in the receiver will be removed at each successive stroke. But as the quantity of air removed at each stroke is only a fraction of the quantity remaining, we can never produce a perfect vacuum, though we might approach as near to it as we pleased if this were the only obstacle.

130. Calculation of the Degree of Exhaustion.—It is easy to calculate the quantity of air left in the receiver after a given number of strokes of the piston. Let V be the volume of the cylinder, V' that of the receiver, and M the mass of air in the receiver at first. On raising the piston, the air which occupied the volume V' occupies a volume V'+V; of the air thus expanded the volume V is removed, and the volume V' left, being $\frac{V'}{V'+V}$ of the whole quantity or mass M. The quantity remaining after the second stroke is $\frac{V'}{V'+V}$ of that after the first, or is $\left(\frac{V'}{V'+V}\right)^2 M$; and after n strokes $\left(\frac{V'}{V'+V}\right)^n M$. Hence the density and (by Boyle's law) the pressure are each reduced by n strokes to $\left(\frac{V'}{V'+V}\right)^n$ of their original values.

We see, then, that the pressure goes on decreasing indefinitely, and that, consequently, the elasticity of the air may, theoretically at least, be rendered less than any assigned quantity.

131. Mercurial Gauge.—In order to follow the steps of the operation, and to observe at each instant the elastic force of the air in the receiver, the instrument is provided with a siphon-barometer, called the mercurial gauge, inclosed in a bell-shaped vessel of glass F, and communicating by a stop-cock with the receiver. This barometer consists of a bent tube, the branches of which are about a foot in length; one of these is closed and filled with mercury, the other is open. When the pressure of the air in the receiver becomes less than that represented by a column of mercury equal in length to the closed branch of the gauge, the mercury falls, and the elastic force of the air at any moment is given by the difference of level of the

mercury in the two branches; this difference can be measured on a
graduated scale. The mercurial gauge serves to show whether the
instrument is working properly; for instance, in the case of air
getting in anywhere, this would be shown by the fluctuations of the
mercurial column. It also shows when the greatest possible effect
has been attained, by the level of the mercury remaining stationary
notwithstanding the motion of the piston. In theory, as we have
said above, there is no limit to the action of the machine, and at each
stroke of the piston the elastic force of the air should decrease; but
in reality this is not the case, on account of the inevitable imperfec-
tions of the apparatus; there is always a limit, extending further in
proportion to the excellence of the machine, and the barometer shows
the moment when this limit is reached. Instead of a siphon-barometer, we might have an ordinary barometer in connection with
the receiver, and thus observe the progress of the vacuum from the
first strokes of the piston.

132. Admission Stop-cock.—After the receiver has been exhausted
of air, if it was required to raise it from the plate, a very considerable
force would be necessary, amounting to as many times fifteen pounds
as the area of the plate contained square inches. It would, therefore,
be in general impossible to raise the receiver. This is, however, rendered possible by means of the stop-cock R, which is shown in section
above. It is perforated by a straight channel, which, when the
machine is being worked, forms part of the communicating passage.
At 90° from the extremities of this channel is another opening O, forming the mouth of a bent passage, leading to the external air. When we
wish to admit the air into the receiver, we have only to turn the stop-
cock so as to bring the opening O to the side next the receiver; if,
on the contrary, we turn it towards the pump-barrel, all communication between the pump and the receiver is stopped, the risk of air
entering is diminished, and the vacuum remains good for a greater
length of time. This precaution is taken when we wish to leave
bodies in a vacuum for a considerable time. Another method is to
employ a separate plate, which can be detached so as to leave the
machine available for other purposes.

133. Double-barrelled Air-pump.—The machine just described has
only a single pump-barrel; air-pumps of this kind are sometimes
employed, and are usually worked by a lever like a pump-handle.
With this arrangement, it is evidently necessary that the piston, after
having ascended, should descend again to expel the air from the

pump-barrel, and it is only after this double stroke that the operation
can begin anew.

Double-barrelled pumps are more frequently used. An idea of
their general arrangement may be formed from Figs. 136, 137, and
138. Fig. 138 gives the machine in perspective, Fig. 136 is a
section through the axes of the pump-barrels, and Fig. 137 shows
the manner in which communication is established between the
receiver and the two bar-
rels. It will be observed
that the two passages from
the barrels unite in a single
passage to the centre of the
plate p.

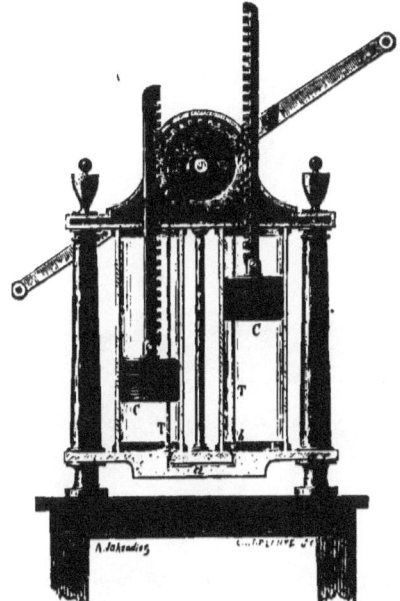

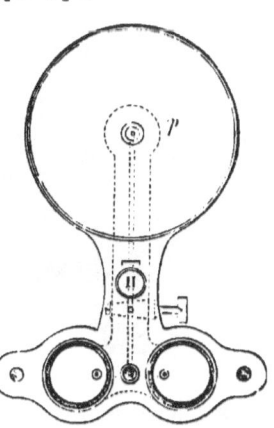

Fig. 136.   Double-barrelled Air-pump.   Fig. 137.

The piston-rods C are two racks working with the pinion P. This
pinion is turned by a double-handed lever, which is worked alternately in opposite directions. In this arrangement, when one piston
ascends the other descends, and consequently in each single stroke
the air of the receiver passes into one or other pump-barrel. A
vacuum is thus produced by half the number of strokes which would
be required with a single-barrelled pump. It has besides another
advantage. In the single-barrelled pump the force required to raise
the piston increases as the exhaustion proceeds, and when it is nearly
completed there is the resistance of almost an atmosphere to be overcome; that is, nearly 15 pounds to the square inch. In the double-

barrelled pump, at the moment when one piston is at the top, and
the other at the bottom, the force opposing the ascent of the one is
precisely equal to that assisting the descent of the other.   We must

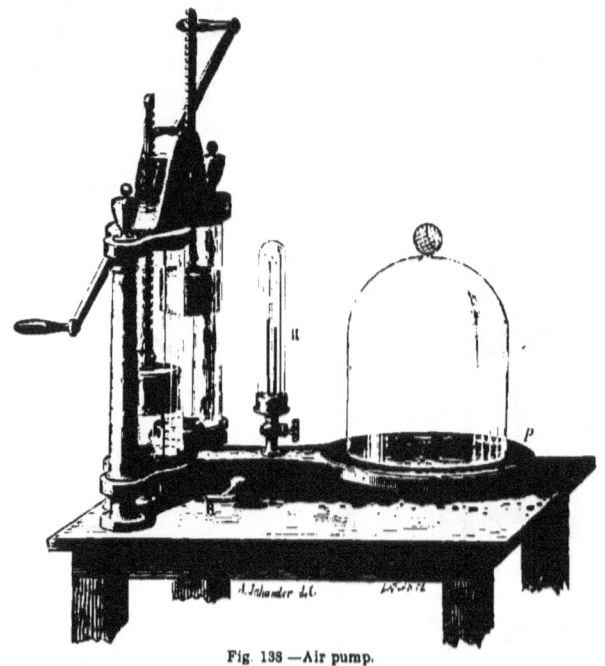

Fig. 138 —Air pump.

observe, however, that this equality exists only at the beginning of
the stroke; for when one of the pistons descends, the air below it is
compressed, its tension becoming greater and greater, until it reaches
that of the atmosphere and raises the piston-valve.   At this moment
the resistance to the ascent of the other piston is entirely uncompen-
sated, and up to this point the compensation has been gradually
diminishing.   But the more nearly we approach to a perfect vacuum,
the more slowly does the tension of the air compressed beneath the
piston increase, so that, unlike the single-barrelled pump, it becomes
easier to work as the exhaustion proceeds.

134. Single-barrelled Pumps with Double Action. — We do not,
however, require two pump-barrels in order to obtain double action,
as the same effect may be obtained with a single barrel.   An arrangement for this purpose was long ago suggested by Delahire for water-

pumps; but the principle has only lately been applied to the construction of air-pumps.

Fig. 139 represents the single barrel of the double-acting pump of Bianchi. It will be seen that the piston-valve opens into the hollow piston-rod; a second valve, also opening upward, is placed at the top of the pump-barrel. Two other openings, one above, the other below, serve to establish communication by means of a bent vertical tube between the pump-barrel and the passage to the plate. These openings are closed alternately by two conical stoppers at the two extremities of a metal rod passing through the piston, and carried with it in its vertical movement by means of friction. When the piston ascends, as in the figure, the upper opening is closed and the lower one is open; when the piston begins to descend, the opposite effect is immediately produced. Accordingly we see that, whichever be the direction in which the piston is moving, the receiver is being exhausted of air. In fact, when

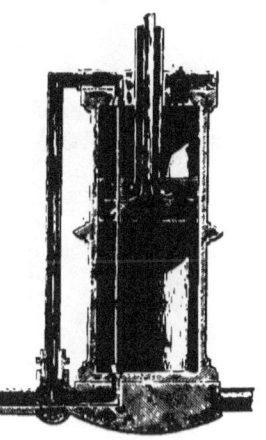

Fig. 139.
Barrel of Bianchi's Air-pump.

the piston ascends, air from the receiver will enter by the lower opening, and the air above the piston will be gradually compressed, and will finally escape by the valve above. In the descending movement, air will enter by the upper opening, and the compressed air beneath the piston will escape by the piston-valve. The movement of the piston is produced by a peculiar arrangement shown in Fig. 140, which gives a general view of the apparatus.

The pump-barrel, which is composed entirely of cast-iron, oscillates about an axis passing through its base. On the top are guides in which the end of a crank travels. The pump is worked by turning a heavy fly-wheel of cast-iron, on the axis of which is a pinion which drives a toothed wheel on the axis of the crank. The end of the crank is attached to the extremity of the piston-rod. It is evident that on turning the fly-wheel the pump-barrel will oscillate from side to side, following the motions of the crank, and the piston will alternately ascend and descend in the barrel, the length of which should be equal to the diameter of the circle described by the end of the crank.

Fig. 140.—Bianchi's Air-pump.

135. Various Experiments with the Air-pump.—At the time when the air-pump was invented, several experiments were devised to show the effects of a vacuum, some of which have become classical, and are usually repeated in courses of experimental physics.

Experiment of the Burst Bladder.—On the plate of an air-pump is placed a glass cylinder open at the bottom, and having a piece of bladder tightly stretched over the top. As the exhaustion proceeds, the bladder bends inwards in consequence of the atmospheric pressure above it, and finally bursts with a loud report.

It often happens that, notwithstanding the strong exterior pressure, the bladder does not give way, its molecules preserving their equilibrium of cohesion. But this equilibrium is, so to speak, unstable, and a few taps are sufficient to destroy it and cause the bladder to burst.

Magdeburg Hemispheres. —We take two hemispheres (Fig 142), which can be exactly fitted on each other; their exact adjustment is further assisted by a projecting internal rim, which is smeared with lard. The apparatus is exhausted of air

Fig. 141.
Burst Bladder.

Fig. 142.
Magdeburg Hemispheres.

through the medium of the stop-cock attached to one of the hemispheres, and when a vacuum has been produced, it will be found that a considerable force is required to separate the two parts, and this force increases with the size of the hemispheres.

This resistance to a force of separation is due to the normal exterior pressure of the air on every point of the surface, a pressure which is counterbalanced by only a very feeble pressure from the interior. In order to estimate the resultant effect of these different pressures, let us suppose that the external surface, instead of being spherical, is formed of a series of steps; that is to say, of alternate vertical and horizontal elements. It is evident that the pressures exerted upon these latter will have no influence upon the adhesion of the hemispheres; the first alone produce this effect, and the sum of these is evidently equal to the pressure of the atmosphere upon the circular area forming the common base of the hemispheres. For example, if this area is ten square inches, the hemispheres will be pressed against each other with a force of 150 pounds.

Fountain in Vacuo.—The apparatus for this experiment consists of a bell-shaped vessel of glass (Fig. 143), the base of which is pierced

Fig.143.—Fountain in Vacuo.

by a tube fitted with a stop-cock which enables us to exhaust the vessel of air. If, after a vacuum has been produced, we place the lower end of the tube in a vessel of water, and open the stop-cock, the liquid, being pressed externally by the atmosphere, mounts up the tube and ascends in a jet into the interior of the vessel. This experiment is often made in the opposite manner. Under the receiver of the air-pump is placed a vial partly filled with water, and having its cork pierced by a tube open at both ends, the lower end being beneath the surface of the water. As the exhaustion proceeds, the air in the vial, by its excess of pressure, acts upon the liquid and makes it issue in a jet.

136. Limit to the Action of the Air-pump.—We have said above (§ 131) that the air-pump does not continue the process of rarefaction indefinitely, but that at a certain stage its effect ceases, and the tension of the air in the receiver undergoes no further diminution. If the pump is very badly made, this tension is considerable; but even with the most perfect machines it is always sensible. A pump such as we have described may be considered as very good if it reduces the tension of the air in the receiver to one-fiftieth of an inch of mercury: it is very rarely that a lower limit is reached.

LEAKAGE.—This limit to the action of the machine is due to various causes. One of these is evidently the leakage at different parts of the apparatus. It is impossible to prevent the air from getting in at several points; and although at the beginning of the operation the quantity of air which thus enters is small in comparison with that which is-pumped out, still, as the exhaustion proceeds, the air enters faster, on account of the diminished internal pressure, and at the same time the quantity expelled at each stroke becomes less, so that at length a point is reached at which the inflow and outflow are equal.

SPACE UNTRAVERSED BY PISTON.—Another reason of imperfect exhaustion is that, after all possible precautions, a space is still left between the bottom of the pump-barrel and the lower surface of the piston when the latter is at the end of its downward stroke. It is evident that at this moment the air contained in this *untraversed space* is of the same tension as the atmosphere. On raising the piston, this air is indeed rarefied; but it still preserves a certain tension, and it is evident that when the air in the receiver has been brought to this stage of rarefaction, the machine will cease to produce any effect.

If v is the volume of this space, V the volume of the pump-barrel, the air, which at volume v has a tension H equal to that of the atmosphere, will have, at volume V, a tension equal to $H\frac{v}{V}$. This gives the limit to the action of the machine as deduced from the consideration of the untraversed space.

AIR GIVEN OUT BY OIL.—Finally, perhaps the most important cause, and the most difficult to remedy, is the absorption of air by the oil used for lubricating the pistons. This oil is poured on the top of the piston, but the pressure of the external air forces it between the piston and the barrel, whence it falls in greater or less quantity to the bottom of the barrel, where it absorbs air, and partially yields it up at the moment when the piston begins to rise, thus evidently tending to derange the working of the machine. It has been attempted to get rid of untraversed space by employing a kind of piston of mercury. This has also the advantage of fitting the barrel more accurately, and thus preventing the entrance of air. The use of oil is at the same time avoided, and we thus escape the injurious effects mentioned above. We proceed to describe two machines founded upon this principle.

137. Kravogl's Air-pump.—This contains a hollow glass cylinder AB tapering at the upper end, and surmounted by a kind of funnel. The piston is of the same shape as the cylinder, and is covered with a layer of mercury, whose depth over the point of the piston is about

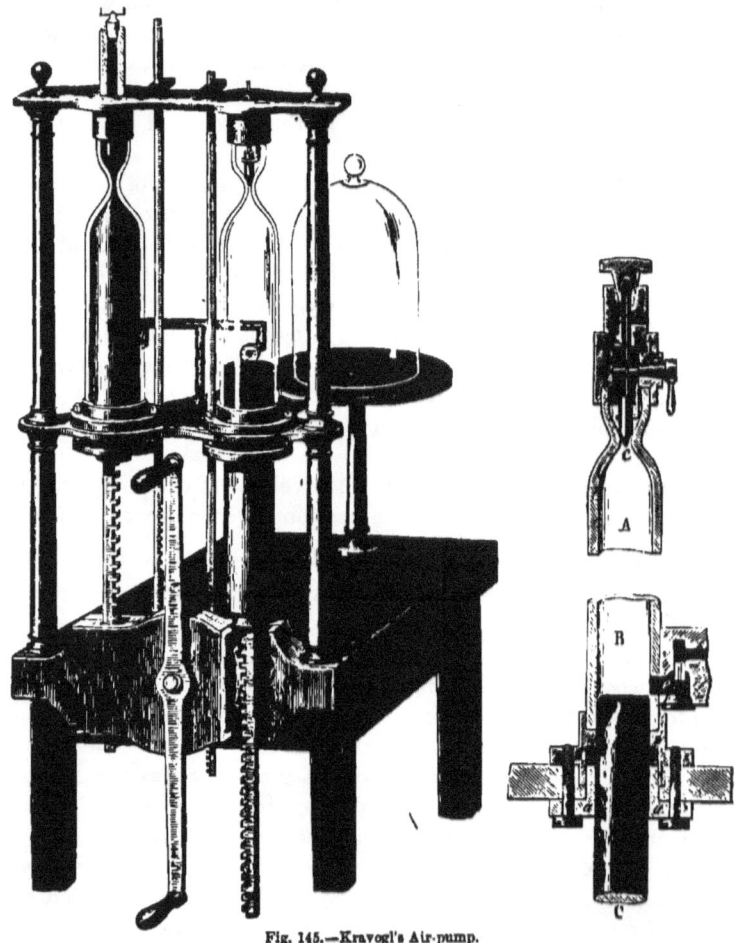

Fig. 145.—Kravogl's Air-pump.

$\frac{1}{10}$th of an inch when the piston is at the bottom of its stroke, but is nearly an inch when the piston rises and fills the funnel-shaped cavity in which the pump-barrel terminates. A small interval, filled by the liquid, is left between the barrel and the piston; but at the bottom of the barrel the piston passes through a leather box carefully made, so as to be perfectly air tight.

The air from the receiver passes through the lateral opening e; it is driven before the mercury into the funnel above. With the air passes a certain quantity of mercury, which is detained by a steel valve c at the narrowest part of the funnel. This valve rises automatically when the surface of the mercury is at a distance of about half an inch from the funnel, and falls back into its former position when the piston is at the end of its upward stroke. In the downward stroke, when the mercury is again half an inch from the funnel, the valve opens again and allows a portion of the mercury to pass.

The effect of this arrangement is easily understood; there is no "untraversed space," the presence of the mercury above and around the piston causes a very complete fit, and excludes the external air; and hence the machine, when well made, is very effective.

When this is the case, and when the mercury used in the apparatus is perfectly dry, a vacuum of about $\frac{1}{250}$th of an inch can be obtained. The dryness of the mercury is a very important condition, for at ordinary temperatures the elastic force of the vapour of water has a very sensible value. If we wish to employ the full powers of the machine, we must have, between the vessel to be exhausted of air and the pump-barrel, a desiccating apparatus.

The arrangement of the valve e is peculiar. It is of a conical form, so as, in its lowest position, to permit the passage of air coming from the receiver. Its ascent is produced by the pressure of the mercury, which forces it against the conical extremity of the passage, and the liquid is thus prevented from escaping.

The figure represents a double-barrelled machine analogous to the ordinary air-pump. Besides the pinion working with the racks of the pistons, there is a second smaller pinion, not shown in the figure, which governs the movements of the valves c. All the parts of this machine, as the stop-cocks, valves, pipes, &c., must be of steel, to avoid the action which the mercury would have upon any other metal.

138. Geissler's Machine.—Geissler, of Bonn, has invented a mercurial air-pump, in which the vacuum is produced by communication of the receiver with the Torricellian vacuum. Fig. 146 represents this machine as constructed by Alvergniat. It consists of a vertical tube, which serves as a barometric tube, and communicates at the bottom, by means of a caoutchouc tube, with a globe which serves as the cistern.

At the top of the tube is a three-way stop-cock, by which communication can be established either with the receiver to the left, or with a funnel to the right, which latter has an ordinary stop-cock at the bottom. By means of another stop-cock on the left, communication with the receiver can be opened or closed. These stop-cocks are made entirely of glass. The machine works in the following manner: communication being established with the funnel, the globe which serves as cistern is raised, and placed, as shown in the figure, at a higher level than the stop-cock of the funnel. By the law of equilibrium in communicating vessels, the mercury fills the barometric tube, the neck of the funnel, and part of the funnel itself. If the communication between the funnel and tube be now stopped, and the globe lowered, a Torricellian vacuum is produced in the upper part of the vertical tube.

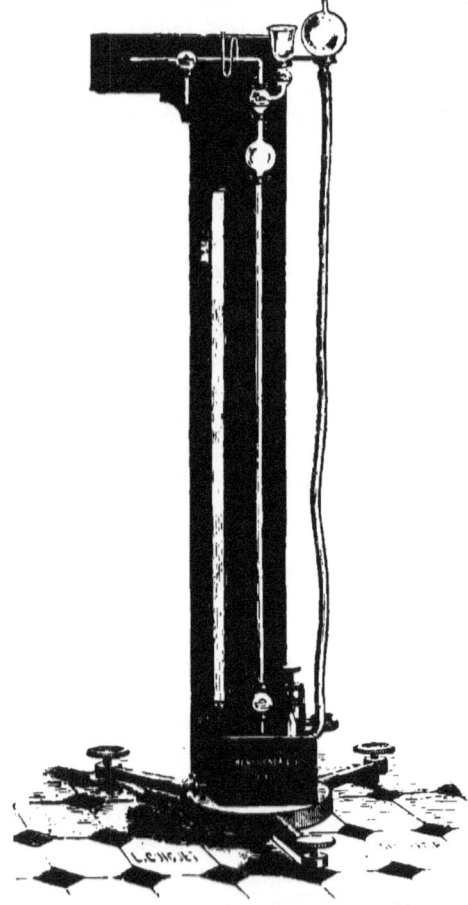

Fig. 146.—Geissler's Machine.

Communication is now opened with the receiver; the air rushes into the vacuum, and the column of mercury falls a little. Communication is now stopped between the tube and receiver, and opened between the tube and the funnel, the simple stop-cock of the funnel being, however, left shut. If at this moment the globe is replaced in the position shown in the figure, the air endeavours to escape by the funnel, and it is easy to allow it to do so. Thus, a part of the air of the receiver has been removed, and the apparatus is in the same position as at the beginning. The operation described is equivalent to a stroke of the piston in the ordinary machine, and this process must be repeated till the receiver is exhausted.

As the only mechanical parts of this machine are glass stop-cocks, which are now executed with great perfection, it is capable of giving very good results. With dry mercury a vacuum of $\frac{1}{250}$th of an inch may very easily be obtained. The working of the machine, however, is inconvenient, and becomes exceedingly laborious when the receiver is large. It is therefore employed directly only for producing a vacuum in very small vessels; when the spaces to be exhausted of air are at all large, the operation is begun with the ordinary machine, and the mercurial air-pump is only employed to render the vacuum thus obtained more perfect.

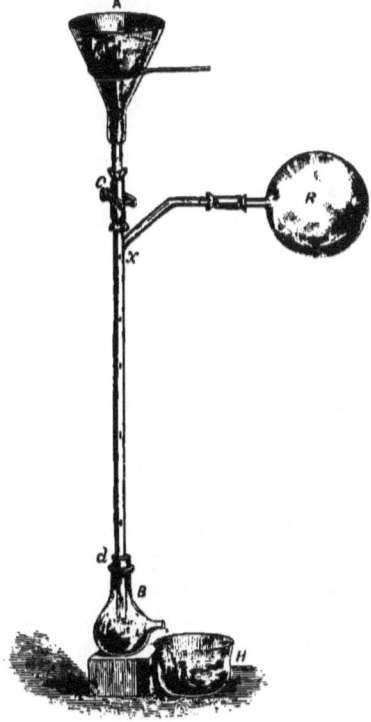

138 A. Sprengel's Air-pump.— This instrument, which may be regarded as an improvement upon Geissler's, is represented in its simplest form in Fig. 146 A. cd is a glass tube longer than a barometer tube, down which mercury is allowed to fall from the funnel A. Its lower end dips into the glass vessel B, into which it is fixed by means of a cork. This vessel has a spout at its side, a few millimetres higher than the lower end of the tube. The first portions of mercury which

Fig. 146 A.—Sprengel's Air-pump.

run down will consequently close the tube, and prevent the possibility of air entering it from below. The upper part of cd branches off at x into a lateral tube communicating with the receiver R, which it is required to exhaust. A convenient height for the whole instrument is six feet. The funnel A is supported by a ring as shown in the figure, or by a board with a hole cut in it. The tube cd consists of two parts, connected by a piece of india-rubber tubing, which can be compressed by a clamp so as to keep the tube closed when desired. As soon as the mercury is allowed to run down the exhaustion begins, and the whole length of the tube, from x to d, is seen to be filled with cylinders of mercury separated by cylinders of air, all moving downwards. Air and mercury escape through the spout of the bulb B, which is above the basin H, where the mercury is collected. This has to be poured back from time to time into the funnel A, to pass through the tube again and again until the exhaustion is completed.

As the exhaustion is progressing, it will be noticed that the inclosed air between the mercury cylinders becomes less and less, until the lower part of cd presents the aspect of a continuous column of mercury about 30 inches high. Towards this stage of the operation a considerable noise begins to be heard, similar to that of a shaken water-hammer, and common to all liquids shaken in a vacuum. The operation may be considered completed when the column of mercury does not inclose any air, and when a drop of mercury falls upon the top of this column without inclosing the slightest air-bubble. The height of this column now corresponds exactly with the height of the column of mercury in a barometer; or, what is the same, it represents a barometer whose vacuum is the receiver R and connecting tube.

Dr. Sprengel recommends the employment of an auxiliary air-pump of the ordinary kind, to commence the exhaustion when time is an object, as without this from 20 to 30 minutes are required to exhaust a receiver of the capacity of half a litre. As, however, the employment of the auxiliary pump involves additional connections and increased leakage, it should be avoided when the best possible exhaustion is desired. The fall tube must not exceed about a tenth of an inch in diameter, and special precautions must be employed to make the india-rubber connections air-tight. (See *Chemical Journal* for 1865, p. 9.)

By this instrument air has been reduced to $\frac{1}{1500000}$th of atmospheric density, and the average exhaustion attainable by its use is about one-millionth, which is equivalent to ·00003 of an inch of mercury.

139. Double Exhaustion.—In the mercurial machines just described there is no "untraversed space," as the liquid completely expels all the air from the pump-barreL These machines are of very recent invention. Babinet long before introduced an arrangement for the purpose, not of getting rid of this space, but of exhausting it of air.

For this purpose, when the machine ceases to work with the ordinary arrangement, the communication of the receiver with one of the pump-barrels is shut off, and this barrel is employed to exhaust the air from the other. This change is effected by means of a stop-cock at the point of junction of the passages leading from the two barrels (Fig. 147). The stop-cock has a T-shaped aperture, the point of intersection of the two branches being in constant communication with the receiver. In a differ-

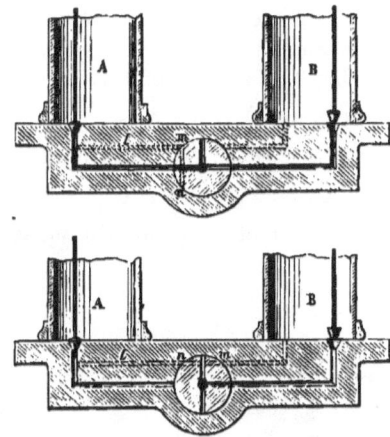

ent plane from that of the T-shaped aperture is another aperture mn, which, by means of the tube l, establishes communication between the pump-barrel B and the communicating passage of the pump-barrel A. From this explanation it will be seen that if the stop-cock be turned as shown in the first figure, the two pump-barrels both communicate with the receiver, and the operation proceeds in the ordinary manner. But if the stop-cock be turned through a quarter of a revolu-

Fig. 147.—Babinet's Doubly-exhausting Stop-cock.

tion, as shown in the second figure, the pump-barrel B alone communicates with the receiver, while it is itself exhausted of air by the barrel A.

It is easy to express by a formula the effect of this double exhaustion. Suppose the pump to have ceased, under the ordinary method of working, to produce any farther exhaustion, the air in the receiver has therefore reached a tension nearly equal to $H\frac{v}{V}$ (§ 136). At this moment the stop-cock is turned into its second position. When the piston B descends, the piston A rises, and the air of the "untraversed space" in B is drawn into A and rarefied. During the inverse operation the air in A is prevented from returning to B, and thus the rarefied air from B, becoming still further rarefied, will draw a fresh quantity of air from the receiver. This air will then be driven into A, where it will be compressed by the descending movement of the piston, and will find its way into the air outside.[1]

This double exhaustion will itself cease to work when air ceases to pass from the pump-barrel B into the pump-barrel A. Now when the piston in this latter is raised, the elastic force of the air which was contained in its "untraversed space" is equal to $H\frac{v}{V}$, for on the last opening of the valve, the air in this space escaped into the atmosphere. On the other hand, when the piston in B is at the end of its upward stroke, the tension of the air is the same as in the receiver. Let this be denoted by x. When the piston in B descends the air is compressed into the "untraversed space" and the passage leading to A. Let the volume of this passage be l. Then the tension will increase, and become $x\dfrac{V+l}{v+l}$ When the machine ceases to produce any farther effect, this tension cannot be greater than that in the pump-barrel A, which is $H\frac{v}{V}$; we have thus, to determine the limit to the action of the pump, the equation

$$x\frac{V+l}{v+l} = H\frac{v}{V}, \text{ whence}$$

$$x = H \cdot \frac{v}{V} \cdot \frac{v+l}{V+l}.$$

140. Air-pump with Free Piston.—We shall describe one more air-pump, constructed by Deleuil, and founded upon an interesting principle. We know that gases possess a remarkable power of adhesion for solids, so that a body placed in the atmosphere may be considered as covered with a very thin coat of air, forming, so to speak, a permanent envelope. On account of this circumstance, gases find very great difficulty in moving in very narrow spaces. On these facts depends the principle of what is called the air-pump with free piston.

The piston P (Fig. 149), which is composed entirely of metal, is of a considerable length, and on its outer surface is a series of parallel circular grooves very close together. It does not touch the pump-barrel at any point; but the distance between the two is very small, about ·001 of an inch. This free piston is surrounded by a cushion of gas, which forms its only stuffing, and is sufficient to enable the

---

[1] It will be observed that during the process of double exhaustion the piston of B behaves like a solid piston; its valve never opens, because the pressure below it is always less than atmospheric.

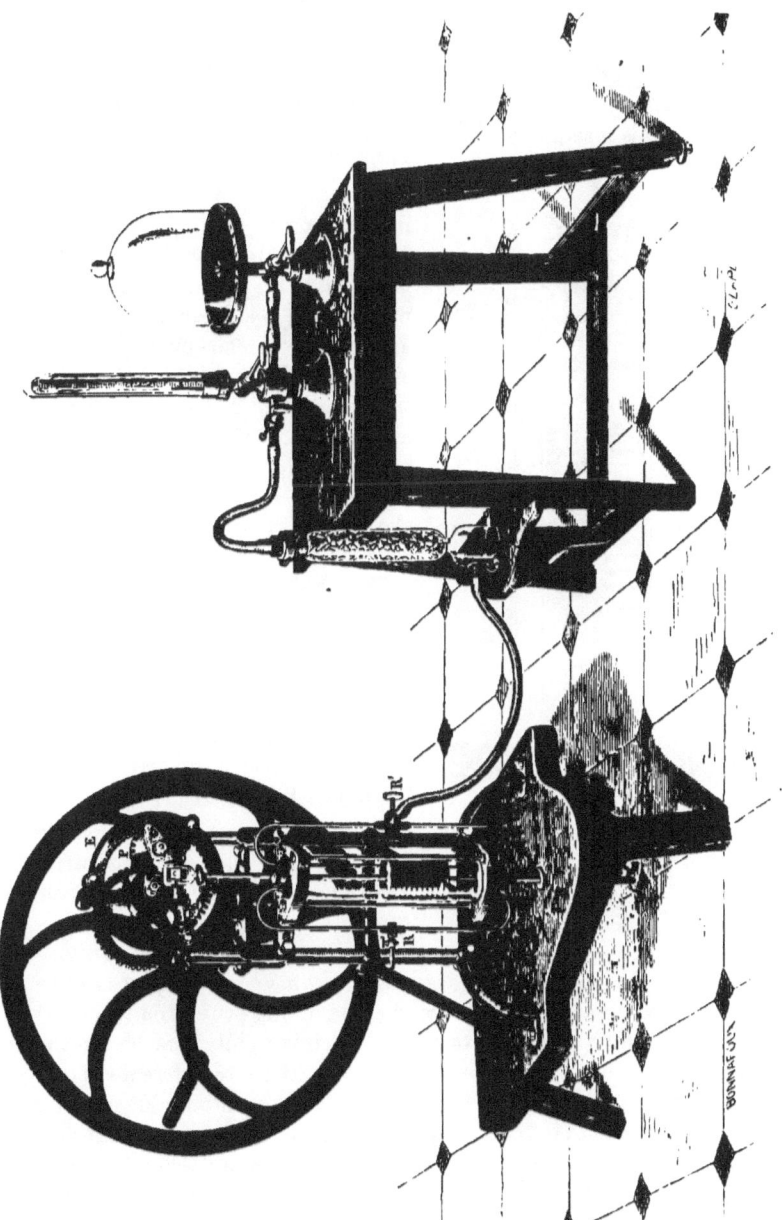

Fig. 148.—Delenil's Air-pump.

machine to work in the ordinary manner, notwithstanding the permanent communication between the upper and lower surfaces of the piston. This machine gives a vacuum about as good as is obtainable

Fig. 149.
Piston and Barrel of Deleuil's Air-pump.

by ordinary pumps, and it has the important advantages of not requiring oil, and of having less friction. It consequently wears better, and is less liable to the development of heat, which is a frequent source of annoyance in air-pumps. It is single-barrelled with double action, like Bianchi's. The two openings S and S' are to admit air from the receiver; they are closed and opened alternately by conical stoppers at the ends of the rod T, which passes through the piston, and is carried with it by friction in its movement. They communicate with tubes which unite at R' with a tube leading from the receiver. A and A' are valves for the expulsion of the air, which escapes by tubes uniting at R. The alternate movement of the piston is produced by what is called Delahire's gearing. This depends on the principle, that *when a circle rolls without sliding in the interior of another circle of double the diameter, any point on the circumference of the rolling circle describes a diameter of the fixed circle.* In order to utilize this property, the end of the piston-rod is jointed to the extremity of a piece of metal which is rigidly attached to the pinion P, the joint being exactly opposite the circumference of the pinion. This latter is driven by a fly-wheel with suitable gearing, and works with the fixed wheel E, which is toothed on the inside. Thus the piston will freely, and without any lateral effort, describe a vertical line, the length of the stroke being equal to the diameter of the fixed wheel.

141. Condensing Pump.—It can easily be seen from the description of the air-pump, that if the expulsion-valves were connected with a tube communicating with a reservoir, the air removed by the pump would be forced into this reservoir. This communication is established in the instrument just described. If, therefore, R' be made to communicate with the external air, this air will be continually drawn in at that point and forced back into the reservoir connected with R, so that the instrument will act as a condensing pump.

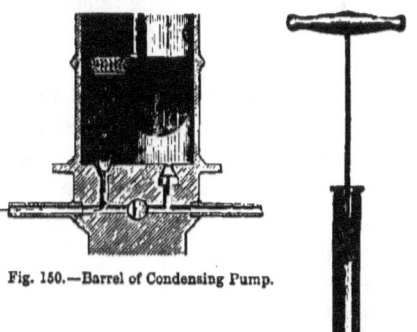

Fig. 150.—Barrel of Condensing Pump.

Fig. 151.
Condensing Pump.

The condensing pump is thus seen to be the same instrument as the air-pump, the only difference being that the receiver is connected with the expulsion-valves, instead of with the exhaustion-valves; it is thus, so to speak, the air-pump reversed.

This fact can be very well seen in the structure of a small pump frequently employed in the laboratory, and represented in Fig. 150.

At the bottom of the pump-barrel are two valves, communicating with two separate reservoirs, that on the left being an admission-valve, and that on the right an expulsion-valve.

When the piston is now raised, rarefaction is produced in the reservoir to the left; and when it is pushed down, the air in the reservoir to the right is compressed.

142. In Fig. 151 is represented a condensing pump often employed. At the bottom of the pump-barrel is a valve *b* opening downward; in a lateral tube is an admission-valve *a* opening inward. The position of these valves is shown in the figure. They are conical metal stoppers, fitted with a rod passing through a hole in a small plate behind, an arrangement which prevents the valve from overturning. The rod is surrounded by a small spiral spring, which keeps the valve pressed against the opening. If now the lower part of the pump-barrel be screwed upon a reservoir, at each upward stroke of

the piston the barrel will be filled with air through the valve a, and at every downward stroke this air will be forced into the reservoir.

If the lateral tube be made to communicate with a bladder or gas-holder filled with any gas, this gas will be forced into the reservoir, and compressed.

143. Calculation of the Effect of the Instrument.—The density of the compressed air after a given number of strokes of the piston may easily be calculated. If v be the volume of the pump-barrel, and V that of the reservoir; at each stroke of the piston there is forced into the reservoir a volume of air equal to that of the pump barrel, which gives a volume nv at the end of n strokes. The air in the reservoir, accordingly, which when at atmospheric pressure had density D, and occupied a volume $V + nv$, will, when the volume is reduced to V, have the density $D\frac{V+nv}{V}$, and the pressure will, by Boyle's law, be $\frac{V+nv}{V}$ atmospheres.

If this formula were rigorously applicable in all cases, there would be no limits to the pressure attainable, except those depending on the strength of the reservoir and the motive power available.

But, in fact, the untraversed space left below the piston when at the end of its downward stroke, sets a limit to the action of the instrument, just as in the common air-pump. For when the air in the barrel is reduced from the volume of the barrel v to that of the untraversed space v', its tension becomes $H\frac{v}{v'}$, and this air cannot pass into the reservoir unless the tension of the air in the reservoir is less than this quantity. This is accordingly the utmost limit of compression that can be attained.

We must, however, carefully distinguish between the effects of untraversed space in the air-pump and in the compression-pump. In the first of these instruments the object aimed at is to rarefy the air to as great a degree as possible, and untraversed space must consequently be regarded as a defect of the most serious importance.

The object of the condensing pump, on the contrary, is to compress the air, not indefinitely, but up to a certain point. Thus, for instance, one pump is intended to give a compression of five atmospheres, another of ten, &c. In each of these cases the maker provides that this limit shall be reached, and accordingly the untraversed space can have no injurious effect beyond increasing the number of strokes required to produce the desired amount of condensation.

144. Various Contrivances for producing Compression.—In order to expedite the process of compression, several pumps such as we have described are combined, which may be done in various ways. Fig. 152 represents the system employed by Regnault in his investigations connected with Boyle's law and the elastic force of vapour. It consists of three pumps, the piston-rods of which are jointed to

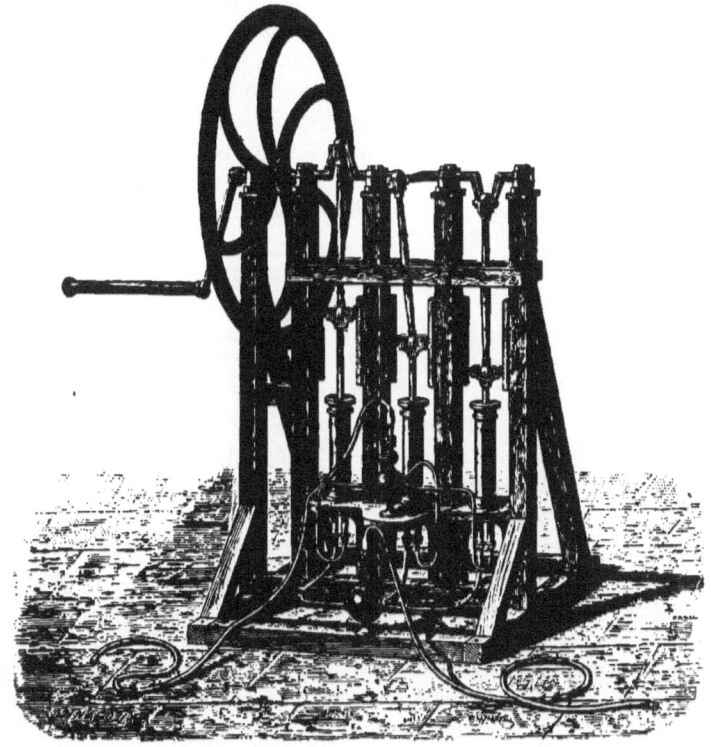

Fig. 152.—Connected Pumps.

three cranks on a horizontal axle, by means of three connecting-rods. This axle, which carries a fly-wheel, is turned by means of one or two handles. The different admission-valves are in communication with a single reservoir in connection with the external air, and the compressed gas is forced into another reservoir which is in communication with the experimental apparatus.

A serious obstacle to the working of these instruments is the heat generated by the compression of the air, which expands the different parts of the instrument unequally, and often renders the piston so tight that it can scarcely be driven. In some of these instruments which are employed in the arts, this inconvenience is lessened by keeping the lower valves covered with water, which has the additional advantage of getting rid of "untraversed space." In this way a pressure of forty atmospheres may easily be obtained with air. Air may also be compressed directly, without the intervention of pumps, when a sufficient height of water can be obtained. It is only necessary to lead the liquid in a tube to the bottom of a reservoir containing air. This air will be compressed until its tension exceeds that of the atmosphere by the amount due to the height of the summit of the tube. It is by a contrivance of this kind that the compressed air is obtained which drives the boring-machines employed in the tunnel through Mont Cenis.

145. Practical Applications of the Air-pump and of Compressed Air. —Besides the use made of the air-pump and the compression-pump in the laboratory, these instruments are variously employed in the arts.

The air-pump is employed by sugar-refiners to lower the boiling point of the syrup. Compression-pumps are used by soda-water manufacturers to force the carbonic acid into the reservoirs containing the water which is to be aerated. The small apparatus described above (Fig. 151) is sufficient for this purpose: it is only necessary to fill the side-vessel with carbonic acid, and to pour a certain quantity of water into the reservoir below. Compressed air has for several years been employed to assist in laying the foundations of bridges in rivers where the sandy nature of the soil requires very deep excavations. Large tubes called *caissons*, in connection with a condensing pump, are gradually let down into the river; the air by its pressure keeps out the water, and the workmen, who are admitted into the apparatus by a sort of lock, are thus enabled to walk on dry ground.

In pneumatic despatch tubes, which have recently been established in many places, a kind of train is employed, consisting of a piston preceded by boxes containing the despatches. By exhausting the air at the forward end of the tube, or forcing in compressed air at the other end, the train is blown through the tube with great velocity.

The atmospheric railway, which was for a few years in existence, was worked upon the same principle: an air-tight piston travelled through a fixed tube, and was connected by an ingenious arrangement with the train above.

Excavating machines driven by compressed air are coming into extensive use in mining operations. They have the advantage of assisting ventilation, inasmuch as the compressed air, which at each stroke of the machine escapes into the air of the mine, cools as it expands.

In the air-gun the bullet is projected by a portion of compressed air which, on pulling the trigger, escapes into the barrel from a reservoir in which it has been artificially compressed.

We may add that the large machines employed in iron-works for supplying air to the furnaces, are really compression-pumps.

CHAPTER XVI.

146. The Baroscope.—Atmospheric air exerts, as we have already mentioned (§ 101), an upward pressure on bodies surrounded by it. This pressure, according to the principle of Archimedes, which applies to gases as well as to liquids, is equal to the weight of the air displaced. Hence it follows that the weight of a body in the air is not its actual weight, but differs from it by a quantity equal to the upward pressure on the body. This principle is illustrated by the baroscope.

This is a kind of balance, the beam of which supports two balls of very unequal sizes, which balance each other in the air. If the

Fig. 153.— Baroscope.

apparatus is placed under the receiver of an air-pump, after a few strokes of the piston, the beam will be seen to incline towards the larger ball, and the inclination will increase as the exhaustion proceeds. The reason is that the air, before it was pumped out, produced an upward pressure, which is now removed. The weight of each ball is thus increased by that of an equal volume of air, whose density is the difference between the densities at the beginning and end of the experiment. This addition is greater for the larger ball, which therefore preponderates.

If after exhausting the air, carbonic acid, which is heavier than air, were allowed to enter the receiver, the large ball would be subjected to a greater increase of upward pressure than the small one, and the beam would incline to the side of the latter.

147. Balloons.—Suppose a body to be lighter than an equal volume of air, then this body will rise in the atmosphere. For example, if we fill soap-bubbles with hydrogen, and shake them off from the end of the tube at which they are formed, they will be seen, if sufficiently large, to ascend in the air. This curious experiment is due to the philosopher Cavallo, who announced it in 1782.[1]

The same principle applies to balloons, which may essentially be reduced to an envelope inclosing a gas lighter than air. In consequence of this difference of density, we can always, by taking a

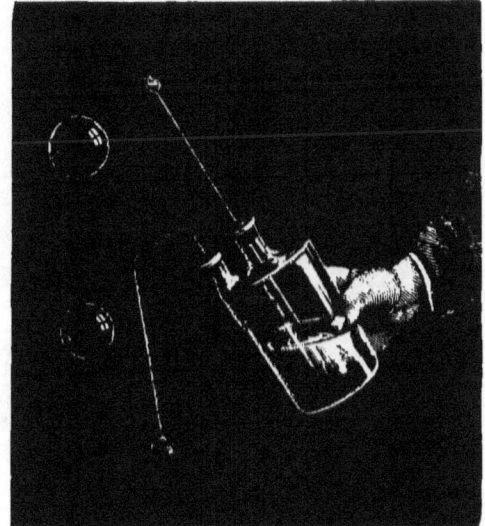

Fig. 154.—Ascent of Soap-bubbles filled with Hydrogen.

sufficiently large volume, make the weight of the gas and containing envelope less than that of the air displaced. In this case the balloon will ascend.

The invention of balloons is due to the brothers Joseph and Stephen Montgolfier. The balloons made by them were globe-shaped, and

[1] The first idea of a balloon must be attributed to Francisco de Lana, who, about 1670, proposed to exhaust the air in globes of copper of sufficient size and thinness to weigh less, under these conditions, than the air displaced. The experiment was not tried, and would certainly not have succeeded, for the pressure of the atmosphere would have caused the globes to collapse. The theory, however, was thoroughly understood by the author, who made an exact calculation of the amount of force tending to make the globes ascend. —D.

constructed of paper, or of paper covered with cloth, the air inside being rarefied by the action of heat. It is curious to remark that in their first attempts they employed hydrogen gas, and showed that balloons filled with this gas could ascend. But as the hydrogen readily escaped through the paper, the flight of the balloons was short, and thus the use of hydrogen was abandoned, and hot air was alone employed.

The name *montgolfières* is still applied in France to fire-balloons. They generally consist of an envelope with a wide opening below,

Fig. 155.—Fire-balloon of Pilatre de Rozier.

under which is hung a brazier,[1] in which, at the moment of ascent, combustibles are placed, and the ascending power of the balloon is thus kept up for some time.

The first public experiment of the ascent of a balloon was performed at Annonay on the 5th June, 1783. On October 21st of the same year, Pilatre de Rozier and the Marquis d'Arlandes achieved the first aerial voyage in a fire-balloon, represented in our figure.

Charles proposed to reintroduce the use of hydrogen by employing

[1] A sponge, dipped in spirits of wine, and ignited, is frequently employed as the source of heat, and is fixed in its place by a light wire-frame.

an envelope impermeable to the gas. This is usually made of silk varnished on both sides, or of two sheets of silk with a sheet of india-rubber between. Instead of hydrogen, coal-gas is now generally employed, on account of its cheapness and of the facility with which it can be procured.

148. Buoyancy.—The buoyancy or lifting power of a balloon is the difference between its weight and that of the air displaced. It is easy to compare the three modes of inflation with respect to the buoyancy which they respectively afford.

A cubic metre of air weighs..............................1·300 grammes.
A cubic metre of hydrogen............................... ·089 ,,
A cubic metre of coal-gas ....about ·750 ,,
A cubic metre of air heated to 200° Cent............. ·800 ,,

We thus see that the buoyancy per cubic metre with hydrogen is 1·211, with coal-gas ·550, and with hot air about ·500 grammes. If, for instance, the total weight to be raised is estimated at 1500 grammes, the volume of a balloon filled with hydrogen capable of raising the weight will be $\frac{1500}{1·210}=1239$ cubic metres. If coal-gas were employed the required volume would be $\frac{1500}{·550}=2727$ cubic metres.

The car in which the aeronauts sit is usually made of wicker-work or whalebone. It is sustained by cords attached to a net-work covering the entire upper half of the balloon, so as to distribute the weight as evenly as possible. The balloon terminates below in a kind of neck opening freely into the air. At the top there is another opening in the inside, which is closed by a valve held to by a spring. Attached to the valve is a cord which passes through the interior of the balloon, and hangs above the car within reach of the hand of the aeronaut.

When the aeronaut wishes to descend, he opens the valve for a few moments and allows some of the gas to escape. An important part of the equipment consists of sand-bags for ballast, which are gradually emptied to check too rapid descent. In the figure is represented a contrivance called a parachute, by means of which the descent is sometimes effected. This is a kind of large umbrella with a hole at the top, from the circumference of which hang cords sup-porting a small car. When the parachute is left to itself, it opens out, and the resistance of the air, acting upon a large surface, moderates the rate of descent. The hole at the top is essential to safety, as it affords a regular passage for air which would otherwise escape from time to time from under the edge of the parachute, thus producing oscillations which might prove fatal to the aeronaut.

One very important precaution to be observed is not to inflate the

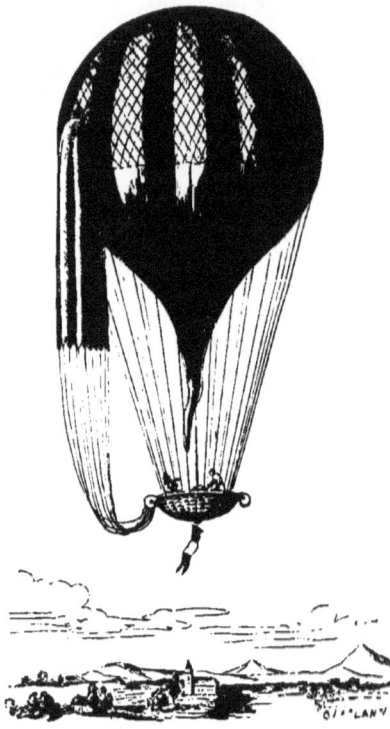

Fig. 156.—Balloon with Car and Parachute.

balloon completely at the commencement of the ascent. The reason is, that the atmospheric pressure diminishing as the balloon ascends, the expansive power of the gas contained produces an increasing effect, as in the experiment described in § 18, and the result would probably be the bursting of the balloon. As the balloon ascends, it increases in volume; but until it is completely inflated, the buoyancy remains constant. Suppose, for instance, that the atmospheric pressure is reduced by one-half, the volume of the balloon will be doubled; it will thus displace a volume of air twice as great as before, but the density of this air will only be half as much, so that the buoyancy remains the same. This conclusion, however, is not quite exact, because part of the balloon, as the cords, the car, &c., are of invariable volume; the density of the air displaced by them is constantly diminishing, and consequently the buoyancy diminishes also. If the balloon continues to rise after it is completely inflated, its buoyancy diminishes rapidly, becoming zero when a stratum of air is reached in which the weight of the volume displaced is equal to that of the balloon itself. It is carried past this stratum in the first instance in virtue of the velocity which it has acquired, and finally comes to rest in it after a number of oscillations.

149. Theory of the Balloon.—The tension of the air in this stratum, the radius of the balloon, and the weight of the different parts, are

connected by a relation which can be very easily established, if we neglect variations of temperature, and which may serve as a guide in the construction of the balloon. If V be the volume of the balloon in litres, the weight of air displaced in grammes is[1]

$$1\cdot293 \; \mathrm{V}\frac{h}{\mathrm{H}},$$

h being the pressure in the stratum of equilibrium, and H that at the surface of the earth. If w and v be the weight and volume of the solid parts, including the aeronauts themselves, δ the density of the gas in the balloon, the equation of equilibrium will be $(\mathrm{V}+v) \; 1\cdot293 \; \frac{h}{\mathrm{H}}$

$$= \mathrm{V}\cdot\delta\frac{h}{\mathrm{H}}+w.$$

In this equation w and v include the weight and volume of the substance composing the balloon and net-work, and therefore are not altogether independent of V, the volume of the balloon. The equation is thus in reality rather complicated, but it may be solved approximately by trial, or by known algebraic methods, and we may find from it the size of balloon necessary for reaching a stratum of a given pressure; and by § 112 we can find the height corresponding to this pressure.

150. **Effect of the Air upon the Weight of Bodies.**—The upward pressure of the air impairs the exactness of weights obtained even with a perfectly true balance, tending, by the principle of the baroscope, to make the denser of two equal masses preponderate. The stamped weights used in weighing are, strictly speaking, standards of mass, and will equilibrate any equal masses in vacuo; but in air the equilibrium will be destroyed by the greater upward pressure of the air upon the larger and less dense body. When the specific gravities of the weights and of the body weighed are known, it is easy from the apparent weight to deduce the true weight (that is to say, the mass) of the body.

Let x be the real weight of a body which balances a weight marked P pounds.

The apparent weight of the body is $x-\frac{x}{\mathrm{D}} \, a = x\left(1-\frac{a}{\mathrm{D}}\right)$, D being the density of the body, and a that of air.[2] The body marked as

---

[1] The weight of a litre (or cubic decimetre = 61·027 cubic inches) of dry air at temperature 0° Cent., and pressure of 760 millimetres of mercury, is 1·293 grammes. Reduced to British measure, this gives 32·7 grains as the weight of 100 cubic inches.

[2] The value of a varies according to the temperature and pressure (see Chap. xxiii.) Its ordinary value is about $\frac{1}{770}$. (See § 100.)

weighing P pounds has in air the apparent weight $P - \frac{P}{M}a = P$ $\left(1 - \frac{a}{M}\right)$, M denoting the density of the substance of which it is composed. These two apparent weights balance one another; hence we have the equation

$$x\left(1 - \frac{a}{D}\right) = P\left(1 - \frac{a}{M}\right),$$

whence

$$x = P\frac{1 - \frac{a}{M}}{1 - \frac{a}{D}} = P\left\{1 + a\left(\frac{1}{D} - \frac{1}{M}\right)\right\} \text{ nearly.}$$

Let us take, for instance, a piece of sulphur whose weight has been found to be 100 grains, the weights being of copper, the density of which is 8·8. The density of sulphur is 2.

We have, by applying the formula,

$$x = 100\left\{1 + \frac{1}{770}\left(\frac{1}{2} - \frac{1}{8·8}\right)\right\} = 100·05 \text{ grains.}$$

We see then that the difference is not altogether insensible. It varies in sign, as the formula shows, according as M or D is the greater. When the density of the body to be weighed is less than that of the weights used, the real weight is greater than the apparent weight; if the contrary, the case is reversed. If the body to be weighed were of the same density as the weights used, the real and apparent weights would be equal. We may remark, that in determining the ratio of the weights of two bodies of the same density, we need not concern ourselves with the effect of the upward pressure of the air, as the correcting factor, which has the same value for both cases, will disappear in the quotient.

CHAPTER XVII.

151. Machines for raising water have been known from very early ages, and the invention of the common pump is pretty generally ascribed to Ctesibius, teacher of the celebrated Hero of Alexandria; but the true theory of its action was not understood till the time of Galileo and Torricelli.

152. Reason of the Rising of Water in Pumps.—Suppose we take a tube with a piston at the bottom, and immerse the lower end of it in water. The raising of the piston tends to produce a vacuum below it, and the atmospheric pressure, acting upon the external surface of the liquid, compels it to rise in the tube and follow the upward motion of the piston. This upward movement of the water would take place even if some air were interposed between the piston and the water; for on raising the piston this air would be rarefied, and its pressure no longer balancing that of the atmosphere, this latter pressure would cause the liquid to ascend in a column, whose weight, added to the pressure of the air below the piston, would be equal to the atmospheric pressure. This is the principle on which water rises in pumps. These instruments have a considerable variety of forms, of which we shall describe the most important types.

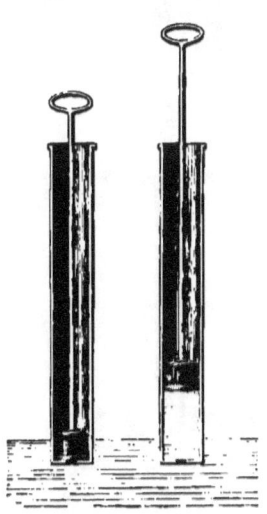

Fig. 157.—Principle of Suction pump.

153. Suction-pump.—The suction-pump consists of a cylindrical pump-barrel traversed by a piston, and communicating by means of a smaller tube, called the suction-tube, with the water in the pump-well. At the junction of the pump-barrel and the tube is a valve opening upward, called the suction-valve, and in the piston is an opening closed by another valve, also opening upward.

Suppose now the suction-tube to be filled with air at the atmospheric pressure, and the water consequently to be at the same level inside the tube and in the well. Suppose the piston to be at the end of its downward stroke, and to be now raised. This motion tends to produce a vacuum below the piston, hence the air contained in the suction-tube will open the suction-valve, and rush into the pump-barrel. Its elastic force being thus diminished, the atmospheric pressure will cause the water to rise in the tube to a height such that the pressure due to this height, increased by the pressure of the air inside, will exactly counter-balance the pressure of the atmosphere. If the piston now descends, the suction-valve closes, the water remains at the level to which it has been raised, and the air, being compressed in the barrel, opens the piston-valve and escapes. At the next stroke of the piston the water will rise still further, and a fresh portion of air will escape.

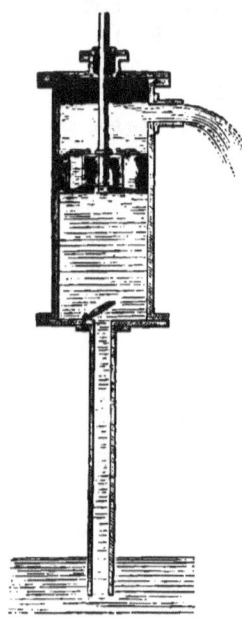

Fig. 158.—Suction-pump.

If, then, the length of the suction-tube is less than about 30 feet, the water will, after a certain number of strokes of the piston, be able to reach the suction-valve and rise into the pump-barrel. When this point has been reached the action changes. The piston in its downward stroke compresses the air, which escapes through it, but the water also passes through, so that the piston when at the bottom of the pump-barrel will have above it all the water which has previously risen into the barrel. If the piston be now raised, supposing the total height to which it is raised to be not more than 34 feet above the level of the water in the well, as should always be the case, the water will follow it in its upward movement, and will fill the pump-barrel. In the downward stroke this water will be forced through the piston-valve, and in the following upward stroke will be discharged at the spout. A fresh quantity of water will by this time have risen into the pump-barrel, and the same operations will be repeated.

We thus see that from the time when the water has entered the pump-barrel, at each upward stroke of the piston a volume of water is ejected equal to the contents of the pump-barrel.

In order that the water may be able to rise into the pump-barrel, the suction-valve must not be more than 34 feet above the level of the water in the well, otherwise the water would stop at a certain point of the tube, and could not be raised higher by any farther motion of the piston.

Moreover, in order that the working of the pump may be such as we have described, that is, that at each upward stroke of the piston a quantity of water may be removed equal to the volume of the pump-barrel, it is necessary that the piston when at the top of its stroke should not be more than 34 feet above the water in the well.

154. Condition that the Water may reach the Pump-barrel.—If the piston does not descend to the bottom of the barrel, the water may fall short of rising to the suction-valve, even though the total height reached by the piston be less than 34 feet. For, if the piston when at the end of its downward stroke leaves below it a space containing air, the tension which this air possesses when the piston is raised diminishes by a corresponding quantity the height to which the water can attain. If, for instance, the length of the suction-tube is 33·5 feet, and the tension of the air remaining above it is, when at its least value, equal to the pressure of 1 foot of water, it is evident that the total height to which the water can rise will be less than 33 feet, and it will, in consequence, be unable to reach the pump-barrel.

Example. The suction-valve of a pump is at a height of 27 feet above the surface of the water, and the piston, the entire length of whose stroke is 7·8 inches, when at the lowest point is 3·1 inches from the fixed valve; find whether the water will be able to rise into the pump-barrel.

When the piston is at the end of its downward stroke, the air which it leaves below it is at the atmospheric pressure; when the piston is raised this air becomes rarefied, and its pressure, by Boyle's law, becomes $\frac{3·1}{10·9}$ that of the atmosphere; this pressure can therefore balance a column of water whose height is $34 \times \frac{3·1}{10·9}$ feet, or 9·67 feet. Hence, the maximum height to which the water can attain is 34 —

9·67 feet = 24·33 feet; and consequently, as the suction-tube is 27 feet long, the water will not rise into the pump-barrel, even supposing the pump to be perfectly free from leakage.

Practically, the pump-barrel should not be more than about 25 feet above the surface of the water in the well; but the spout may be more than 34 feet above the barrel, as the water after rising above the piston is simply pushed up by the latter, an operation which is independent of atmospheric pressure. Pumps in which the spout is at a great height above the barrel are commonly called *lift-pumps*, but they are not essentially different from the suction-pump.

155. Force necessary to raise the Piston.—The force which must be expended in order to raise the piston, is equal to the weight of a column of water, whose base is the section of the piston, and whose height is that to which the water is raised. Let S be the section of the piston, P the atmospheric pressure upon this area, h the height of the column of water which is above the piston in its present position, and h' the height of the column of water below it; then the upper surface of the piston is subjected to a pressure equal to $P + Sh$; the lower face is subjected to a pressure in the opposite direction equal to $P - Sh'$, and the entire downward pressure is represented by the difference between these two, that is, by $S(h + h')$.

The same conclusion would be arrived at even if the water had not yet reached the piston. In this case, let l be the height of the column of water raised; then the pressure below the piston is $P - Sl$; the pressure above is simply the atmospheric pressure P, and, consequently, the difference of these pressures acts downward, and its value is Sl.

156. Efficiency of Pumps.—From the results of last section it would appear that the force required to raise the piston, multiplied by the height through which it is raised, is equal to the weight of water discharged multiplied by the height of the spout above the water in the well. This is an illustration of the principle of work (§ 17A). As this result has been obtained from merely statical considerations, and on the hypothesis of no friction, it presents too favourable a view of the actual efficiency of the pump.

Besides the friction of the solid parts of the mechanism, there is work wasted in generating the velocity with which the fluid, as a whole, is discharged at the spout, and also in producing eddies and other internal motions of the fluid. These eddies are especially produced at the sudden enlargements and contractions of the passages through which the fluid flows. To these drawbacks must be added loss from leakage of water, and at the commencement of the operation from leakage of air, through the valves and at·the circumference of the piston. In common household pumps, which are generally roughly made, the *efficiency* may be as small as ·25 or ·3; that is to say, the product of the weight of water raised, and the height through which it is raised, may be less than the work done in driving the pump in the ratio of one of these numbers to unity.

In Figs. 159 and 160 are shown the means usually employed for working the piston. The first figure needs no explanation; it will be seen that the upward and downward movement of the piston is effected by means of a lever. The second figure represents an arrangement often employed, in which the alternate motion of

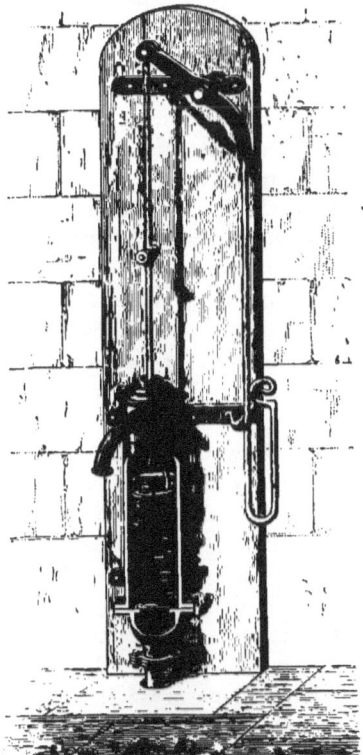

Fig. 159.

Suction-pump.

Fig. 160.

the piston is effected by means of a rotatory motion. For this purpose the piston-rod T is joined by means of the connecting-rod

B to the crank C of an axle turned by a handle attached to the fly-wheel V.

157. Forcing-pump.—The forcing-pump consists of a pump-barrel dipping into water, and having at the bottom a valve opening up-

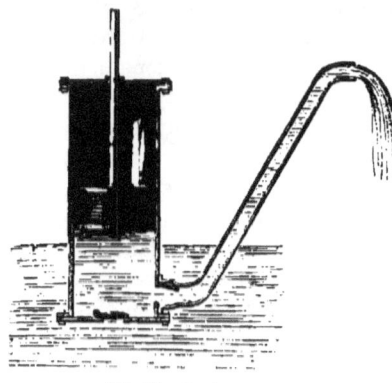

Fig. 161.— Forcing-pump.

ward. In communication with the pump-barrel is a side-tube with a valve at the point of junction opening from the barrel into the tube. A solid piston moves up and down the pump-barrel, and it is evident that when this piston is raised, water enters the barrel by the lower valve, and that when the piston descends, this water is forced into the side-tube. The greater the height of this tube, the greater will be the force required to push the piston down, for the resistance to be over-come is the pressure due to the column of water raised.

The forcing-pump most frequently has a short suction-pipe leading from the reservoir, as represented in Fig. 163. In this case the water is raised from the reservoir into the barrel by atmospheric pressure during the up-stroke, and is forced from the barrel into the ascending pipe in the down-stroke.

158. Plunger.—When the height to which the water is to be forced is very considerable, the different parts of the pump must be very strongly made and fitted together, in order to resist the enormous pressure produced by the column of water, and to prevent leakage. In this case the ordinary piston stuffed with tow or leather washers cannot be used, but is replaced by a solid cylinder of metal called a *plunger.* Fig. 164 represents a section of a pump thus constructed. The plunger is of smaller section than the barrel, and passes through a stuffing-box in which it fits air-tight. The volume of water which enters the barrel at each up-stroke, and is expelled at the down-stroke, is the same as the volume of a length of the plunger equal to the length of stroke; and the hydrostatic pressure to be overcome is pro-portional to the section of the plunger, not to that of the barrel. As the operation proceeds, air is set free from the water, and would even-tually impede the working of the pump were it not permitted to escape.

For this purpose the plunger is pierced with a narrow passage, which is opened from time to time to blow out the air.

The drainage of deep mines is usually effected by a series of pumps. The water is first raised by one pump to a reservoir, into which dips the suction-tube of a second pump, which sends the water up to a second reservoir, and so on. The piston-rods of the different pumps are all joined to a single rod called the *spear*, which receives its motion from a steam-engine.

159. Fire-engine. — The ordinary fire-engine is formed by the union of two forcing-pumps which play into a common reservoir, containing in its upper portion

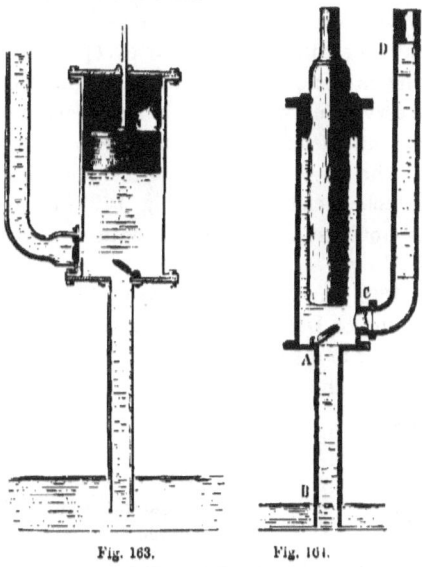

Fig. 163.　　　Fig. 164.
Suction and Force Pump.

(called the air-chamber) air compressed by the working of the engine.

Fig. 162.—Fire-engine.

A tube dips into the water in this reservoir, and to the upper end of this tube is screwed the leather hose through which the water is discharged. The piston-rods are jointed to a lever, the ends of which are raised and depressed alternately, so that one piston is ascending while the other is descending. Water is thus continually being forced into the common reservoir except at the instant of reversing stroke, and as the compressed air in the air-chamber performs the part of a reservoir of work (nearly analogous to the fly-wheel), the discharge of water from the nozzle of the hose is very steady.

The engine is sometimes supplied with water by means of an attached cistern (as in Fig. 162) into which water is poured; but it is more usually furnished with a suction-pipe which renders it self-feeding.

160. Double-acting Pumps.—These pumps, the invention of which is due to Delahire, are often employed for household purposes. They consist of a pump-barrel V V (Fig. 165), with four openings in it, A, A′, B, B′. The openings A and B′ are in communication with the suction-

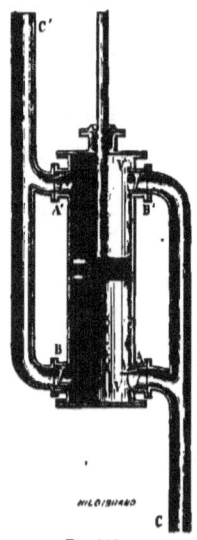

tube C; A′ and B are in communication with the ejection-tube C′. The four openings are fitted with four valves opening all in the same direction, that is, from right to left, whence it follows that A and B′ act as suction-valves, and A′ and B as ejection-valves, and, consequently, in whichever direction the piston may be moving, the suction and ejection of water are taking place at the same time.

161. Rotatory Pump.—Double-acting pumps produce a continuous suction of water. The same end may be attained by means of rotatory pumps, which are largely used in some countries. The figure represents one of these pumps as constructed by Dietz.

The pump-barrel consists of a cylindrical drum B (Fig. 166), containing within it a second cylinder A, of less diameter, and of nearly the same length, open at the ends, which can be turned about its axis by means of a handle; around this

Fig. 165.
Double-action Pump.

axis is a cam or guide mn $m'n'$, rigidly fixed to one of the ends of the drum B. In the box A are four slits, into which fit four blades p; these latter are constantly guided by the movement of one of their ends upon the cam, while the other ends move along the interior surface of the drum B, and the broad iron plate $b'a'$ ab, whose distance from the cam is everywhere equal to the length of one of the blades. By means of two holes in this plate, communication is established between the pump-barrel and the suction-tube C, and between the pump-barrel and the ejection-tube C'. From this arrangement it follows, that on causing the box A to rotate in the direction shown by the arrow, a partial vacuum will be produced behind the tongues, and the water will be drawn in at this side, and ejected at the opposite side.

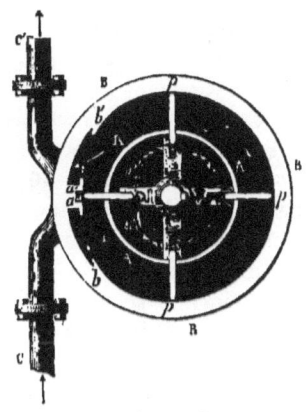

Fig. 166.—Rotatory Pump.

162. **Hydraulic Press.**—The hydraulic press (Fig. 167) consists of a suction and force pump *aa* worked by means of a lever turning about an axis O. The water drawn from the reservoir BB' is forced along the tube CG into the cistern V. In the top of the cistern is an opening through which moves a heavy metal plunger AA. This carries on its

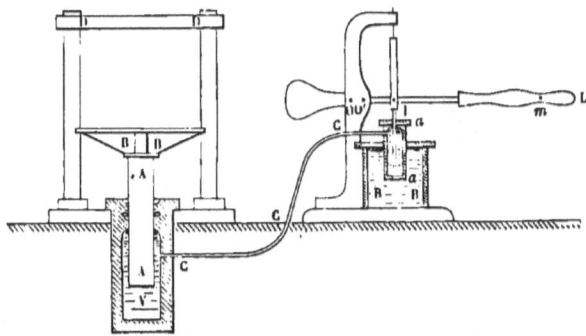

Fig. 167.—Bramah Press.

upper end a large plate BB, upon which are placed the objects to be pressed. Suppose the cistern V to be at first empty, and the piston A to be carried by its own weight to the bottom of the cistern; under these circumstances, suppose the pump to begin to work. The cistern first begins to fill with water; then the pressure exerted by the piston of the pump is transmitted, according to the principles laid down in § 63, to the bottom of the piston A; this piston

accordingly rises, and the objects to be pressed, being intercepted
between the plate and the top of a fixed frame, are subjected to the
transmitted pressure.   The amount of this pressure depends both on
the ratio of the sections of the pistons and on the length of the lever
used to work the force-pump.   Suppose, for instance, that the distance of the point m, where the hand is applied, from the point O, is
equal to twelve times the distance IO, and suppose the force exerted
to be equal to fifty pounds.   By the principle of the lever this is
equivalent to a force of 50×12 at the point I; and if the section of
the piston A be at the same time 100 times that of the piston of the
pump, the pressure transmitted to A will be $50 \times 12 \times 100 = 60,000$
pounds.   These are the ordinary conditions of the press usually employed in workshops.   By drawing out the pin which serves as an
axis at O, and introducing it at O', we can increase the mechanical
advantage of the lever.

Two parts essential to the working of the hydraulic press are not
represented in the figure.   These are a safety-valve, which opens
when the pressure attains the limit which is not to be exceeded; and,
secondly, a tap in the tube C, which is opened when we wish to put
an end to the action of the press.   The water then runs off, and the
piston A descends again to the bottom of the cistern.

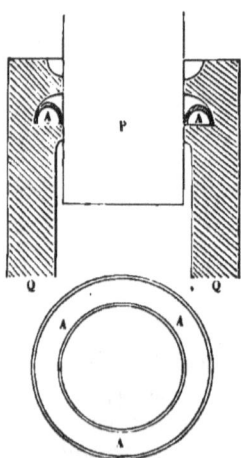

The hydraulic press was clearly described
by Pascal, and at a still earlier date by
Stevinus, but for a long time remained practically useless ; because as soon as the pressure began to be at all strong, the water
escaped at the surface of the piston A.
Bramah invented the *cupped leather collar*,
which prevents the liquid from escaping,
and thus enables us to utilize all the power
of the machine.   It consists of a leather ring
AA (Fig. 168), bent so as to have a semi-
circular section.   This is fitted into a hollow
in the interior of the sides of the cistern, so
that water passing between the piston and
cylinder will fill the concavity of the cupped
leather collar, and by pressing on it will

Fig. 168.—Cup—Leather.

produce a packing that fits more tightly as
the pressure on the piston increases.

The hydraulic press is very extensively employed in the arts.   It is of great power, and may be constructed to give pressures of two or three hundred tons.   It is the instrument generally employed in cases where very great force is required, as in testing anchors or raising very heavy weights.   It was used for raising the sections of the Britannia tubular bridge, and for launching the *Great Eastern.*

CHAPTER XVIII.

163. If an opening is made in the side of a vessel containing water, the liquid escapes with a velocity which is greater as the surface of the liquid in the vessel is higher above the orifice, or to employ the usual phrase, as the *head* of liquid is greater. This point in the dynamics of liquids was made the subject of experiments by Torricelli, and the result arrived at by him was that the velocity of efflux is equal to that which would be acquired by a body falling freely from the upper surface of the liquid to the centre of the orifice. If h be this height, the velocity of efflux is given by the formula

$$V = \sqrt{2\,gh}.$$

This is called Torricelli's theorem; it supposes the sides of the vessel to be thin, and the diameter of the orifice to be very small compared with that of the vessel. It is further assumed that the orifice and the upper surface are under the same conditions as regards atmospheric pressure.

Torricelli's theorem has been regarded as an immediate consequence of the theory of gravitation; according to which, whatever be the path of a heavy body, its velocity depends only on the height of the point of starting above the point finally reached. If this height be h, the velocity is always $\sqrt{2\,gh}$.

But it is not evident that the molecules of a liquid which is escaping are subjected to no force but that of gravity. Besides, the first portions which escape from the vessel do not come from the free surface, and their velocity is due solely to the pressure exerted by the liquid column. It will thus be seen that the velocity of efflux, owing to the complex nature of the phenomenon, can be rigorously established only by the experimental method. It is very easy to perform a simple experiment upon this point. In fact, the molecules issuing from the orifice are ejected with a certain velocity, and should therefore, by the theory of projectiles, describe parabolic paths. The jet issuing from the vessel should accordingly be parabolic, and by measuring its range, we can calculate the velocity of efflux.

The experiment may easily be made by means of the apparatus represented in Fig. 169. It consists of a cylinder in which are a

Fig. 169.—Apparatus for verifying Torricelli's Theorem.

number of equidistant orifices in the same vertical line. A tap placed above the cylinder supplies the vessel with water, and with the help of an overflow-pipe, maintains the liquid at a constant level, which is as much above the highest orifice as each orifice is above that next below it.

The liquid which escapes is received in a trough, the edge of which is graduated. A travelling piece with an index line engraved on it slides along the trough; it carries, as shown in one of the separate figures, a disc pierced with a circular hole, and capable of being turned in any direction about a horizontal axis passing through its centre. In this way the disc can always be placed in such a position that its plane shall be at right angles to the liquid jet, and that the jet shall pass freely and exactly through its centre. The index line then indicates the range of the parabolic jet with considerable precision. This range is reckoned from the vertical plane containing the orifices, and is measured on the horizontal plane passing through the centre of the disc. The distance of this latter plane below the lowest orifice is equal to that between any two consecutive orifices.

The following is the way in which the result of an experiment is estimated. Let b be the height of the orifice above the horizontal plane through the centre of the ring, and let a be the range of the jet. If the liquid molecules were simply falling from a height, b, they would traverse this space in a time given by the formula

$$b = \frac{g t^2}{2}.$$

On the other hand, if they were simply obeying the force of ejection at the orifice, they would, by virtue of their initial velocity x, traverse the distance a in the same time t, whence we have

$$a = x t.$$

Eliminating t between these two equations, we have

$$b = \frac{g a^2}{2 x^2};$$

whence

$$x = a \sqrt{\frac{g}{2 b}}.$$

On comparing this velocity with that given by Torricelli's theorem, there is generally found to be a small difference between them, as is shown in the subjoined table:—

Head.	Jet.		Velocity.		Ratio of actual to theoretical velocity.
	a.	b.	Actual.	Theoretical.	
metres.	metres.	metres.			
2·20	6·28	7·53	6·65	6·70	·993
3·93	4·66	8·45	8·67	8·70	·988
7·17	1·41	6·25	11·67	11·88	·983

164. Intersection of Jets.—If Torricelli's theorem is correct, the

value of x just found ought to be equal to $\sqrt{2gh}$; thus, we have the equation

$$a\sqrt{\frac{g}{2b}} = \sqrt{2gh}, \text{ whence } a^2 = 4bh. \qquad (3)$$

From this a curious result may be deduced: it will be seen that if h and b vary in such a manner that their produce remains constant, the value of a will remain unchanged. This law may easily be verified experimentally. It is only necessary to open at the same time the top and bottom orifices, or the second and fourth; the result will be two jets which will intersect one another at the centre of the ring.

We may remark, however, that the verification of this law does not prove Torricelli's theorem, for the result would be the same, if in (3) the constant 4 were replaced by any other constant; it proves, however, that the velocity is proportional to the square root of the head of water, and not to the head itself, as was formerly believed.

165. Quantity of Liquid Discharged.—It would appear at first sight that Torricelli's theorem could be tested by a very simple and decisive experiment. Suppose the level of the liquid in a vessel to be maintained constant, and the volume of the liquid which escapes through an orifice during a certain time to be measured. This can be compared with the volume calculated *a priori*, by assuming that the quantity discharged in a unit of time is equal to a cylinder whose base is the section of the orifice and height the velocity; so that the quantity which escapes in time t will be given by the formula

$$Q = ts\sqrt{2gh};$$

s being the section of the orifice. Now, in all the experiments which have been performed, when the orifice is a simple perforation in a thin plate, the actual discharge has been found to be less than this, being generally about ·6 of it.

This discrepancy arises from neglecting the fact, that the particles of liquid at the margin of the jet have a converging motion, in consequence of which the jet contracts rapidly for a small distance after issuing from the orifice. Beyond this small distance the contraction is very gradual, depending only on the continued action of gravity. The portion which forms the termination of the rapid contraction, is called the contracted vein, or *vena contracta*, and its section appears to be about ·6 of that of the orifice. If, then, in the above formula we make s denote the section of the contracted vein, we shall obtain results agreeing with experiment.

166. Efflux-tubes.—This explanation is confirmed by the effect of efflux-tubes. These are pipes not exceeding a few inches in length, which are fitted to the holes in the side of the vessel. If, for instance, a tube of cylindrical form is employed, the contraction will be prevented by the adherence of the liquid to the sides of the tube; the section of the jet is then the same as that of the tube. In this case, if the velocity be measured, it will be found to be less than when the orifice is a hole in a thin plate; but, on multiplying the velocity thus obtained by the section of the tube, we shall obtain a result agreeing with the actual discharge.

The apparatus above described enables us to estimate the differences in velocity caused by efflux-tubes. For this purpose a sliding plate is used, with one orifice and two efflux-tubes, one cylindrical, the other conical; by sliding the plate the liquid can be made to flow out of either of these tubes.

167. Efflux through Pipes.—When the liquid, instead of escaping through a short spout, flows through a long tube, the velocity is considerably reduced by the friction of the molecules against each other, and against the sides of the tube. This velocity is also not the same at all points in the same section; it is least where the liquid is in contact with the tube, and is greatest in the centre of the liquid column. When a uniform delivery has been established, the quantity of water which passes in each unit of time is constant, and is the same for all sections; and the average velocity across any section will be obtained by dividing the volume discharged in a second by the sectional area. Many experiments have been performed with the view of determining this velocity in a certain number of particular cases, and certain empirical rules have thus been arrived at. It is difficult to treat the subject rationally, or to establish results that shall be perfectly general.

168. Fountains.—If the lower end of a water-pipe be connected with a mouth-piece pointing vertically upwards, the liquid on issuing from the opening will rise in a vertical jet to a height depending on the velocity of efflux. If there were no resistance, this height would be that *due to the velocity,* namely $\frac{v^2}{2g}$, according to the formulæ of § 38; but this is not actually the case. The friction of the liquid against itself, and the weight of the particles which fall back upon the rest, counteract the tendency to ascend. The effect of this last cause can be somewhat diminished by slightly inclining the jet.

169. Efflux of a Liquid in contact with Confined Air.—When the surface of the liquid is in contact with a volume of air whose pressure varies, the velocity of efflux is affected by this variation. Let A B C D be a closed vessel filled with a liquid as far as M N, and let the space above contain air at the atmospheric pressure. If a small orifice be opened below, the liquid will begin to flow out; but the air above will become gradually rarefied, until at length its pressure, together with that due to the depth of liquid, will be equal to that of the atmosphere; when this happens the liquid will cease to flow,

Fig. 170.

unless the circumstances are such that air-bubbles can enter. Let us see what will be the height of the liquid column in the vessel when the flow ceases. Let $A C = l$, $A M = h$, and let p be the height of a column of the liquid which will balance the pressure of the atmosphere. The air, which at the beginning of the experiment had a volume $l - h$, and a pressure p, will now have a volume $l - x$, and consequently a pressure $p \frac{l-h}{l-x}$; we have then

$$p \frac{l-h}{l-x} + x = p,$$

whence

$$x = \frac{p + l \pm \sqrt{(p+l)^2 - 4 p h}}{2}.$$

The — sign alone can be taken with the radical, since x must be less than l. This case of equilibrium occurs in several well-known experiments.

Pipette.—This is a glass tube (Fig. 171) open at both ends, and terminating below in a small tapering spout. If a certain quantity of water be introduced into the tube, either by aspiration or by direct immersion in water, and if the upper end be closed with the finger, the efflux of the liquid will cease after a few seconds. On admitting the air above, the efflux will begin again, and can again be stopped at pleasure.

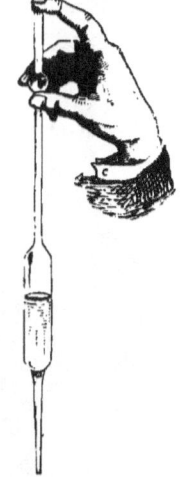

Fig. 171.—Pipette.

The Magic Funnel.—This funnel is double, as is shown in Fig. 172. Near the handle is a small opening by which the space between the two funnels communicates with the external air. Another opening connects this same space with the tube of the inner funnel. If the interval between the two funnels be filled with any liquid,

this liquid will run out or will cease to flow according as the upper
hole is open or closed.   The opening and closing of the hole can be

easily effected with the thumb
of the hand holding the funnel without the knowledge of
the spectator.   This device
has been known from very
early times.

The instrument may be
used in a still more curious
manner.   For this purpose
the space inside is secretly
filled with highly-coloured
wine, which is prevented
from escaping by closing the
opening above.

Fig. 172.—Magic Funnel.

Water is then poured into the central funnel, and escapes either
by itself or mixed with wine, according as the thumb closes or opens

Fig. 173.—Inexhaustible Bottle.

the orifice for the admission of air.   In the second case, the water
being coloured with the wine, it will appear that wine alone is
issuing from the funnel; thus the operator will appear to have the

power of making either water or wine flow from the vessel at his pleasure.

The Inexhaustible Bottle.—The inexhaustible bottle is a toy of the same kind. It is an opaque bottle of sheet-iron or gutta-percha, containing within it five small vials. These communicate with the exterior by five small holes, which can be closed by the five fingers of the hand. Each vial has also a small neck which passes up the large neck of the bottle. The five vials are filled with five different liquids, any one of which can be poured out at pleasure by uncovering the corresponding hole.

170. Intermittent Fountain.—The intermittent fountain is an apparatus analogous to the preceding, except that the interruptions in the efflux are produced auto-matically by the action of the instrument, without the inter-vention of the operator. It consists of a globe V, which can be hermetically closed by means of a stopper, and can be put in communication with efflux-tubes *a*, by which the water contained in it can escape. A vertical tube *t* rises nearly to the top of the globe, and ter-minates below at a short dis-tance from the bottom of the basin B. This basin is pierced with a small opening *o*, by which the water contained in it escapes into the lower basin C. Suppose the globe to be filled with water, and com-munication with the efflux-tubes to be established, then

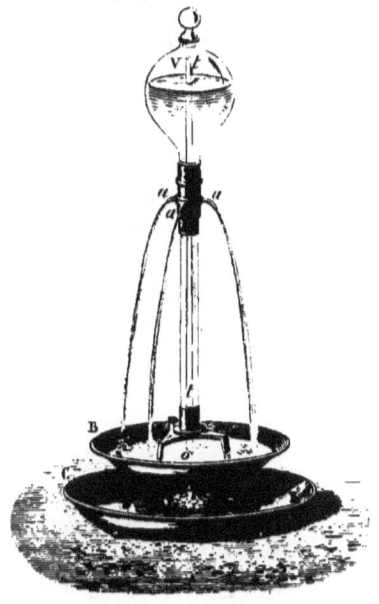

Fig. 174.—Intermittent Fountain.

the liquid will flow into the basin B, and thence into C. But the size of the opening *o* is such that it suffers less water to escape than passes out by the efflux-tubes; the liquid will therefore accumulate in B, and will finally cover the bottom of the tube *t*. Communication between the external air and the upper part of the globe will then be cut off, and the liquid will after a few moments' cease to flow.

But as the basin B continues to be emptied by the opening o, the liquid in the basin will sink below the end of the tube; air will then enter the globe, the liquid will again begin to issue from the efflux-tubes, and so on.

171. Siphon.—The object of the siphon is the transference of liquid from one vessel to another. It essentially consists of a bent tube (Fig. 175) with branches of unequal length. The short branch dips

Fig. 175.—Siphon.

into the liquid to be transferred, the other opens directly into the air. If we suppose the siphon full of liquid, it is easy to see that the liquid will flow from the short to the long branch.

For if we consider (Fig. 176) a layer of liquid M, at the highest point of the siphon, this layer will be subjected to a pressure from left to right equal to the atmospheric pressure diminished by the height DC, or MI.

Let this latter height be h, and let H be the external pressure expressed as the height of an equivalent column of the liquid, then the pressure from left to right will be $H - h$. The pressure from right to left will be $H - EF = H - h'$. Now as h' is greater than h, the first pressure will overcome the second, and the layer M will consequently move from left to right. But if the height of the smaller branch be less than H, the liquid cannot separate, for the pressure of the atmosphere would immediately fill up the vacuum which would be formed. Thus the liquid will continue to flow uninterruptedly until the liquid in the vessel AB has fallen below the level of the end of the shorter branch of the siphon.

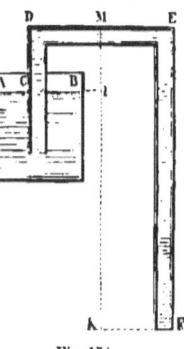

Fig. 176.

The force causing the liquid to flow is the pressure represented by a column of liquid $h' - h$, the velocity of efflux is thus equal to $\sqrt{2g(h'-h)}$, friction being neglected.

In the above reasoning we have supposed the external pressure H to be the same at C and at F. This is evidently the case when the pressure is that of the atmosphere. If we suppose the surrounding medium to be of a density such that the variation of pressure for the levels AB and KF cannot be neglected, the expression for the velocity must be modified. · Let d be the density of the liquid, d' that of the surrounding medium, the excess of pressure from left to right is represented by the weight of a liquid column of density d, and of height $h'-h$, diminished by the weight of a column of the same height of density d'; that is, it is given by the expression $(h'-h)\,d - (h'-h)\,d' = (h'-h)\,(d-d')$. Now the height m of the liquid which would produce the same pressure is given by the equation $md = (h'-h)\,(d-d')$. Thus the velocity of efflux will be

$$V = \sqrt{2\,gm} = \sqrt{\frac{2\,g\,(h-h')\,(d-d')}{d}}.$$

In the case (which could scarcely occur in practice), where d' is greater than d, the pressure from left to right will be negative; that is, the excess of pressure will be from right to left. The liquid will then flow from right to left, and with a velocity given by the above formula if $d-d'$ is replaced by $d'-d$.

172. Starting the Siphon.—In order that the siphon should work, it must first be charged with liquid. This is effected in various ways. When the liquid can be taken into the mouth without danger, the charging can be effected (Fig. 177) by sucking at a side-tube attached to the long branch.

This method is inapplicable to liquids which would have an injurious effect upon the mouth. The following method is often employed in the transfer of sulphuric acid from one vessel to another.

The long branch of the siphon (Fig. 178) is first filled with sulphuric acid. This is effected by means of two funnels (which can be plugged at pleasure) at the bend of the tube. One of these admits the liquid, and the other suffers the air to escape. The two funnels above are then closed, and the tap at the lower end of the tube is opened so as to allow the liquid to escape. The air in the short branch follows the acid, and becomes rarefied; the acid behind it rises, and if it passes the bend, the siphon will be charged; for each portion of the liquid which issues from the tube will draw a corresponding portion from the short to the long branch.

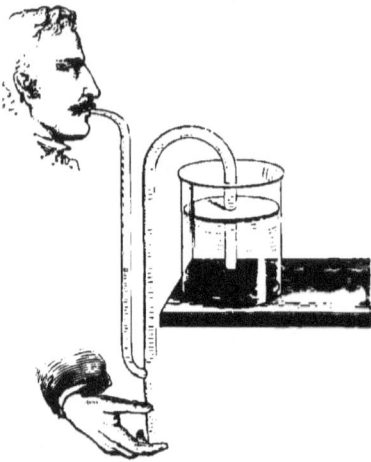

Fig. 177.—Starting the Siphon.

To insure the working of the sulphuric acid siphon, it is not sufficient to have the vertical height of the long branch greater than that of the short branch; it is farther necessary that it should exceed a certain limit, which depends upon the dimensions of the siphon in each particular case. In order to calculate this limit, we must remark that when the liquid begins to flow, its height diminishes in the long and increases in the short branch; if these two heights should become equal, there would be equilibrium.

Fig. 178.—Siphon for Sulphuric Acid.

We see, then, that in order that the siphon may work, it is necessary that when the liquid rises to the bend of the tube, there should be in the long branch a column of liquid whose vertical height is at least equal to that of the short branch, which we shall denote by h, and the actual length of the short branch from the surface of the liquid in which it dips to the summit of the bend by h'. Then if a be the inclination of the long branch to the vertical, and L the length of the long branch, which we suppose barely sufficient, the length of the column of liquid remaining in the long branch will be h sec a. The air which at atmospheric pressure H occupied the length h', now under the pressure $H - h$ occupies a length $L - h$ sec a; hence, by Boyle's law, we have

$$Hh' = (H - h) (L - h \text{ sec } a), \text{ whence } L = h \text{ sec } a + \frac{Hh'}{H - h}.$$

In this formula H denotes the height of a column of sulphuric acid whose pressure equals that of the atmosphere.

173. Vase of Tantalus.—The siphon may be employed to produce the intermittent flow of a liquid. Suppose, for instance, that we

have a vase in which is a bent tube rising to a height n, and with the short branch terminating near the bottom of the vase, while the long branch passes through the bottom. If liquid be poured into the vase, the level will gradually rise in the short branch of the bent tube, and will finally reach and pass the point n, when the siphon will begin to discharge the liquid. If, then, we suppose the liquid to escape by the siphon faster than it is poured into

Fig. 179.—Vase of Tantalus.

the vessel, the level of the liquid will gradually fall below the termination of the shorter branch. The siphon will then empty itself, and will not recommence its action till the liquid has again risen to the level of the bend.

If the cup is made of metal with the siphon concealed in the thickness of the sides, when a person in lifting it to his lips inclines it to the side in which the siphon is, the siphon will become charged, and will empty the vessel. Hence the name *vase of Tantalus* given to this cup in old treatises on physics. Instead of a bent tube we may employ, as in the first figure, a straight tube covered by a bell-glass left open below; in this case the space between the tube and the bell takes the place of the shorter leg of the siphon.

It is to an action of this kind that natural intermittent springs are generally attributed. Suppose a reservoir (Fig. 180) to communicate with an outlet by a bent tube forming a siphon, and suppose it to be fed by a stream of water at a slower rate than the siphon is able to discharge it. When the water has reached the bend, the siphon will become charged, and the reservoir will be emptied; it will then be filled again as far as the bend, and so on.

174. Mariotte's Bottle.—This is an apparatus often employed to obtain a continuous flow of water. It consists of a flask whose cork is pierced by a straight tube open at both ends, and with the lower

Fig. 180.—Intermittent Spring.

extremity descending to a. An efflux-tube is placed at b near the bottom of the flask. Suppose that the flask is full of water, and that the tube is also full to the upper end. If the tube b be now opened, the liquid molecules at the orifice will be pressed inwards with a force equal to the atmospheric pressure, but will be pressed outwards with a force exceeding this pressure by the height of the column of water as far as the upper end of the tube. The liquid will therefore flow out; but no vacuum will be produced in the upper part of the flask, for the pressure of the atmosphere will compel the liquid in the tube to replace that which escapes. The level of the liquid in the tube will thus rapidly fall, and the velocity will gradually decrease, as will be seen by the diminished range of the jet. When the liquid reaches the point a, the efflux will continue; but then air will enter the vase in successive bubbles, and will rise to the upper part of the vase, in such quantity that its pressure, together with that of the height of water above the horizontal plane through a, will maintain a pressure on this plane equal to that of the atmosphere. From this time the liquid will flow with a constant velocity due to the height of a above b. Strictly speaking, inasmuch as the air enters, not in a continuous manner, but in successive bubbles, that is, in jerks, the velocity of discharge oscillates about a constant mean value, but the oscillations are in general almost imperceptible. Instead of the ver-

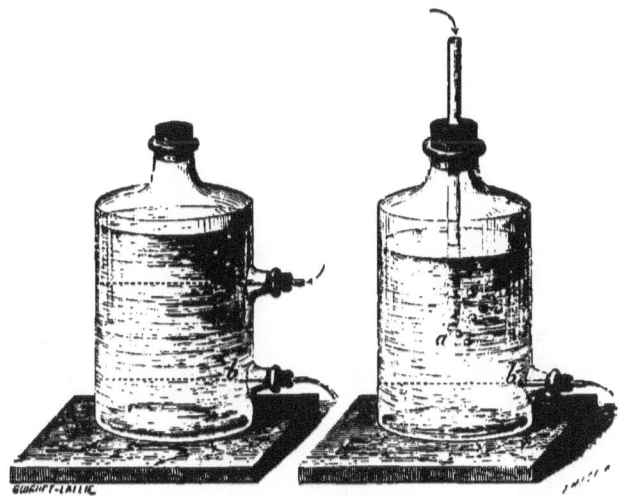

Fig. 181.—Mariotte's Bottle.

tical tube, we may use a vase with two openings at different levels; the liquid escapes by the lower orifice b, while air enters by the upper orifice a. Mariotte's vase is sometimes used in the laboratory to produce the uniform flow of a gas by employing the water which escapes to expel the gas. We may also draw in gas through the tube of Mariotte's bottle; in this case, the flow of the *water* is uniform, but the flow of the *gas* is continually accelerated, since the space occupied by it in the bottle increases uniformly, but the density of the gas in this space continually increases.

www.ingramcontent.com/pod-product-compliance
Lightning Source LLC
Chambersburg PA
CBHW030133060726
47499CB00015B/1642